Understanding the behaviour and improving the welfare of dairy cattle

It is widely recognised that agriculture is a significant contributor to global warming and climate change. Agriculture needs to reduce its environmental impact and adapt to current climate change whilst still feeding a growing population, i.e. become more 'climate-smart'. Burleigh Dodds Science Publishing is playing its part in achieving this by bringing together key research on making the production of the world's most important crops and livestock products more sustainable.

Based on extensive research, our publications specifically target the challenge of climate-smart agriculture. In this way we are using 'smart publishing' to help achieve climate-smart agriculture.

Burleigh Dodds Science Publishing is an independent and innovative publisher delivering high quality customer-focused agricultural science content in both print and online formats for the academic and research communities. Our aim is to build a foundation of knowledge on which researchers can build to meet the challenge of climate-smart agriculture.

For more information about Burleigh Dodds Science Publishing simply call us on +44 (0) 1223 839365, email info@bdspublishing.com or alternatively please visit our website at www.bdspublishing.com.

Related titles:

Achieving sustainable production of milk Volume 3: Dairy herd management and welfare
Print (ISBN 978-1-78676-052-4); Online (ISBN 978-1-78676-054-8, 978-1-78676-055-5)

Improving organic animal farming
Print (ISBN 978-1-78676-180-4); Online (ISBN 978-1-78676-183-5, 978-1-78676-182-8)

Advances in breeding of dairy cattle
Print (ISBN 978-1-78676-296-2); Online (ISBN 978-1-78676-298-6, 978-1-78676-299-3)

Improving rumen function
Print (ISBN 978-1-78676-332-7); Online (ISBN 978-1-78676-334-1, 978-1-78676-335-8)

Improving dairy herd health
Print (ISBN 978-1-78676-467-6); Online (ISBN 978-1-78676-469-0, 978-1-78676-470-6)

Chapters are available individually from our online bookshop:
https://shop.bdspublishing.com

climate•SMART•publishing
IN AGRICULTURAL SCIENCE

BURLEIGH DODDS SERIES IN AGRICULTURAL SCIENCE
NUMBER 98

Understanding the behaviour and improving the welfare of dairy cattle

Edited by Professor Marcia Endres,
University of Minnesota, USA

burleigh dodds
SCIENCE PUBLISHING

Published by Burleigh Dodds Science Publishing Limited
82 High Street, Sawston, Cambridge CB22 3HJ, UK
www.bdspublishing.com

Burleigh Dodds Science Publishing, 1518 Walnut Street, Suite 900, Philadelphia, PA 19102-3406, USA

First published 2021 by Burleigh Dodds Science Publishing Limited

Library of Congress Control Number: 2021931070

British Library Cataloguing in Publication Data
A catalogue record for this book is available from the British Library

ISBN 978-1-78676-459-1 (Print)
ISBN 978-1-78676-462-1 (PDF)
ISBN 978-1-78676-461-4 (ePub)
ISSN 2059-6936 (print)
ISSN 2059-6944 (online)

DOI 10.19103/AS.2020.0084

Typeset by Deanta Global Publishing Services, Dublin, Ireland

Contents

Series list

Title	Series number
Achieving sustainable cultivation of maize - Vol 1 From improved varieties to local applications Edited by: Dr Dave Watson, CGIAR Maize Research Program Manager, CIMMYT, Mexico	001
Achieving sustainable cultivation of maize - Vol 2 Cultivation techniques, pest and disease control Edited by: Dr Dave Watson, CGIAR Maize Research Program Manager, CIMMYT, Mexico	002
Achieving sustainable cultivation of rice - Vol 1 Breeding for higher yield and quality Edited by: Prof. Takuji Sasaki, Tokyo University of Agriculture, Japan	003
Achieving sustainable cultivation of rice - Vol 2 Cultivation, pest and disease management Edited by: Prof. Takuji Sasaki, Tokyo University of Agriculture, Japan	004
Achieving sustainable cultivation of wheat - Vol 1 Breeding, quality traits, pests and diseases Edited by: Prof. Peter Langridge, The University of Adelaide, Australia	005
Achieving sustainable cultivation of wheat - Vol 2 Cultivation techniques Edited by: Prof. Peter Langridge, The University of Adelaide, Australia	006
Achieving sustainable cultivation of tomatoes Edited by: Dr Autar Mattoo, USDA-ARS, USA & Prof. Avtar Handa, Purdue University, USA	007
Achieving sustainable production of milk - Vol 1 Milk composition, genetics and breeding Edited by: Dr Nico van Belzen, International Dairy Federation (IDF), Belgium	008
Achieving sustainable production of milk - Vol 2 Safety, quality and sustainability Edited by: Dr Nico van Belzen, International Dairy Federation (IDF), Belgium	009
Achieving sustainable production of milk - Vol 3 Dairy herd management and welfare Edited by: Prof. John Webster, University of Bristol, UK	010
Ensuring safety and quality in the production of beef - Vol 1 Safety Edited by: Prof. Gary Acuff, Texas A&M University, USA & Prof. James Dickson, Iowa State University, USA	011
Ensuring safety and quality in the production of beef - Vol 2 Quality Edited by: Prof. Michael Dikeman, Kansas State University, USA	012
Achieving sustainable production of poultry meat - Vol 1 Safety, quality and sustainability Edited by: Prof. Steven C. Ricke, University of Arkansas, USA	013
Achieving sustainable production of poultry meat - Vol 2 Breeding and nutrition Edited by: Prof. Todd Applegate, University of Georgia, USA	014

Introduction

The welfare of farmed animals such as cattle is an increasing concern for consumers and regulatory agencies. This collection summarises and reviews the wealth of recent research on understanding dairy cow behaviour and improving their welfare.

Part 1 of this collection reviews advances in understanding aspects of dairy cattle behaviour such as cognition and learning and pain and stress. Part 2 assesses developments in welfare indicators and advances in monitoring such as sensors, acoustic and video techniques. It also examines effective training and certification schemes for improving on-farm dairy cattle welfare. Part 3 summarises recent research on improving welfare practices at different stages in production, including calves, heifers and transition cows, as well as areas such as housing. It also covers optimizing welfare in transport and slaughter of cattle.

Chapter 1 sets the scene by providing the author's overview of factors influencing the sustainability of the dairy sector. Welfare and other aspects of sustainability are becoming increasingly important in consumer purchasing decisions. The chapter introduces several key welfare issues affecting the dairy industry that need some improvement, including lameness, mastitis and aspects of calf management. Other topics mentioned in Chapter 1 that might relate to the sustainability of dairy production include minimising grain use, feeding high protein leaves of shrubs and trees, reducing greenhouse gas emissions such as by changing diet to reduce methane output, and improving labelling and traceability.

Part 1 Understanding behaviour

The first chapter of Part 1 reviews advances in understanding cognition and learning in cattle. Understanding the cognitive capacities of cattle is central when designing husbandry environments and developing handling and management practices. Chapter 2 outlines the current knowledge on cattle learning and cognition, with special emphasis on their socio-cognitive capacities. The chapter begins by discussing learning and memory in cattle, focusing specifically on feature cue learning and memory and spatial learning and memory. It then goes on to discuss the importance of understanding the physical environment, which is followed by a review of social cognition in the cattle group. Sections on human–cow interactions and cognitive research in an applied setting are also provided. The chapter concludes by highlighting the importance of using dairy cattle cognitive capacities in the conceptualisation of handling regimes to ensure a safer working environment for handlers whilst also safeguarding the welfare of the animals.

Chapter 3 focuses on advances in understanding pain and stress in cows. The chapter provides definitions of pain and stress, particularly focusing on the issue that 'stress' as a stand-alone generic term may not be helpful in terms of advancing our understanding of the impact of animal management on animal welfare. For cattle, as for other animals managed by humans, it is necessary to consider the source of 'stress' and the general and specific animal responses to different contexts and events. The chapter also outlines the standard methods of assessing pain and stress, then presents a number of new methods that are being used to assess the experiential aspects. It explains the theoretical background of these methods and how they allow us to gain a better understanding of the experience of pain and stress in cattle than previous methods. Finally, the chapter provides a case study describing a situation where these approaches have led to clear recent improvements in practice that are having real benefits for animals under commercial management.

Part 2 Welfare indicators and monitoring

Part 2 of the book opens with a chapter focused on developing effective welfare measures for cattle. Chapter 4 concentrates on the performance characteristics a welfare measure should possess in order to be considered valid for the assessment of animal welfare. It presents a choice of validation measures that can be used to assess the welfare of cattle and discusses ways they can be collected in practice. The chapter also presents the various definitions of animal welfare and how these definitions can affect the measures that are chosen.

The next chapter provides an overview of advances in precision livestock farming (PLF) techniques for monitoring dairy cattle welfare. Chapter 5 begins by describing the potential of PLF technology linked to the Five Domains framework, then goes into more detail by focusing on the use of PLF techniques for each of the five domains. Finally, the chapter reviews the need for data processing algorithms.

The final chapter of Part 2 discusses developing effective training and certification schemes for improving on-farm dairy cattle welfare. Chapter 6 begins by reviewing the selection of measures that can be used for assessing animal welfare, focusing firstly on animal-based measures, then moves on to review resource and management based measures. The chapter also discusses the four welfare quality principles - feeding, housing, health and behaviour - and how dairy cattle welfare can be affected by the quality of these four principles. Training of assessors is also discussed, before a summary of why using a combination of resource-based, management-based and animal-based measures is the best way to ensure good welfare in animals such as dairy cattle.

Part 3 Improving welfare practices

Part 3 begins with a chapter on developments in housing of cattle to promote health and welfare. Chapter 7 summarises the adaptations and improvements that have been made to freestall housing of adult dairy cattle to enhance health and welfare. The chapter starts by reviewing the impact of housing on natural behaviours of dairy cattle, such as grazing. It then moves on to examine housing adaptations to feeding behaviour which is then followed by a discussion on group management of dairy cattle. A section on promoting optimal resting behaviour is also provided, followed by a review of behaviour responses to extreme climate. The chapter also addresses enhancements to natural behaviour before concluding with a section that emphasises the importance of using well-designed housing facilities to enhance the health and welfare of the animals using them.

The next chapter focuses on advances in understanding behavioural needs and improving the welfare of calves and heifers. Management factors affecting dairy calves have broad impacts on behavioral development and implications for long-term welfare. Chapter 8 summarises recent research addressing behavioral needs of pre-weaned calves and post-weaned heifers, encompassing aspects of social interactions and feeding, rest and comfort, and opportunities for other behavioral expression. In general, evidence suggests welfare and performance benefits of accommodating more natural behavioral expression, including social contact, higher milk intakes fed via a teat, dietary variety, and space and resources to accommodate preferred activities, such as grooming and play. Consequences of restricting these behaviours include development of abnormal oral behaviours. Emerging themes of this area of research include consideration of individual differences and implications and approaches to accommodate broader behavioral expression through increasing environmental complexity.

Moving on from Chapter 8, Chapter 9 addresses advances in understanding the needs and improving the welfare of transition cows. One of the most challenging time periods in a dairy cow's life is when she gives birth and transitions into lactation (the 'transition' period). The chapter begins by focusing on the biological health of dairy cows before and after giving birth, as cows are at high risk of disease post-calving. It then goes on to discuss the affective states of transition dairy cows, including pain associated with labour and disease. Finally, the chapter concludes by assessing the natural maternal behavior of cows before calving through a case study.

The final chapter of the book assesses optimising welfare in transport and slaughter of cattle. Transport represents one of the most stressful events an animal may experience. Its impact on the welfare of beef and dairy cattle is frequently underestimated, as are the effects of transport on carcass quality issues such as bruises and dark cutting beef. Chapter 10 breaks the process down into three phases: pre-transport, transport and arrival at slaughter. The chapter begins by reviewing culling and permanent herd removal, focusing specifically on voluntary versus involuntary culling, surplus replacements and bull sales, culling based upon performance and removal rates for dairy and beef herds. It then discusses transport of cattle to markets and slaughter facilities and the trends in transport of dairy and beef cattle. This is then followed by a review of muscling and the body condition cull cows arriving at slaughter plants. A section on hide and carcass bruising evaluation is also provided, before concluding with a review of optimising welfare during the pre-transport, transport and slaughter stages.

Chapter 1

Dairy cattle welfare and other aspects of sustainability

Donald M. Broom, University of Cambridge, UK

1 Introduction

Future animal production, indeed all kinds of production, will have to be sustainable. A system or procedure is sustainable if it is acceptable now and if its expected future effects are acceptable, particularly in relation to resource availability, consequences of functioning and morality of action (Broom 2014, modified after Broom 2001, 2010). The development of new, sustainable systems is urgently needed because some current livestock production practices are not acceptable to consumers, many of whom now include the ethics of food production in their evaluation of product quality (Broom 2010). The opinion of the public is based on a range of components of sustainability, including welfare (Table 1).

After an era when negative effects on consumer health have been a major factor in attitudes to some foods, poor welfare of animals is now a substantial factor in consumer purchase decisions, which affect the sustainability of animal production systems (Bennett et al. 2002; Broom 2014, 2017a). Recently, production practices negatively affecting the world's climate and biodiversity have also become more important. In the long run, inefficient use of world resources is likely to influence the public greatly but any of these factors can result in a product being avoided by consumers. It is therefore very important for producers to consider all aspects of sustainability, including welfare, if they are to hold on to their markets. The dairy industry is particularly vulnerable because some of its production practices have come under sustained criticism

http://dx.doi.org/10.19103/AS.2020.0084.01

Table 1 Some factors which might make a food production system unsustainable

Inefficient usage of world food resources	
Adverse effects on human welfare, including health	
Poor welfare of animals	
Harmful environmental effects	a too much greenhouse gas production b low biodiversity c insufficient conservation
Unacceptable genetic modification	
Production systems not 'fair trade', producers in poor countries not properly rewarded	
Damage to rural communities	

in recent years. What can it do to meet changing consumer expectations? Since there are many facets of welfare and other aspects of sustainability, many kinds of measurement are needed (Von Keyserlingk et al. 2013).

2 Animal welfare and the dairy industry

Major welfare problems of dairy cows include lameness, mastitis, impaired reproduction, inability to show normal behavioural and physiological responses, infectious diseases and injury (see review by EFSA 2009). Efforts are being made to address these problems but the first three can still be considered a major issue in high-producing dairy cows (Schukken et al. 2005; Oltenacu and Broom 2010; Webster et al. 2017). It would seem that many dairy cows could be at or beyond a production level that is metabolically damaging to them (Knaus 2009). Since some of the major welfare problems of dairy cows are a consequence of genetic selection for high production, there is a need to balance production and functional traits in dairy breeding programmes in order to protect the future of the industry (Van der Werf and Pryce 2019). It is also recommended that cows should not be fed a diet that exacerbates the metabolic disorders leading to lameness, mastitis and reproductive problems (Blümmel et al. 2017). Bovine somatotrophin can further increase milk yield, metabolic pressure and poor welfare resulting from lameness, mastitis and reproductive disorders (EU SCAHAW 1999). This has been known for many years and consumers are increasingly aware of it but its use is still permitted in some countries.

Welfare issues affect both extensive and intensive production systems. None of the problems is necessarily worse in large dairy herds than in small ones but some aspects of welfare, such as outdoor access, may be more challenging to address in large herds (Broom 2013). On the other hand, large herds can have in-house specialised labour (e.g. hoof trimmer and veterinary care), which can improve other aspects of welfare such as animal health (Robbins et al. 2016).

2.1 Lameness

There are many methods for detecting lameness and, as a result, estimates of the prevalence of lameness vary widely. Cook et al. (2016) reported a lameness prevalence of 9.6% in high-performance dairy herds ranging in size from 203 to 2966 cows while others report prevalences of 39% or higher (Archer et al. 2010; Bell 2017). Barker et al. (2010) concluded that a lameness prevalence of 37% was typical of the milk production industry in much of the world and Bell (2017) suggested lameness may affect one-third of adult dairy cattle, with the majority of animals experiencing a lameness event in the course of a year. Lameness can be a significant problem in both intensive and extensive systems (Ranjbar et al. 2016).

Health is an important part of welfare and lameness is a clear example of this. Leg and foot disorders involve painful tissue damage and they are also frustrating because many of them prevent normal control of movement. They often cause animals to walk with a limp, reduce walking to a low level, or avoid walking whenever possible (Broom and Fraser 2015). The ability of lame cattle to carry out various preferred behaviours is generally impaired, and there may be adverse consequences for various other aspects of their normal biological functioning. Lameness reduces ability to obtain some resources, such as food, water and preferred lying places. Lameness always means some degree of poor welfare, and sometimes means that welfare is very poor indeed. EFSA (2009) concluded that no farms should continue to produce milk without urgent remedial action if more than 10% of cows are lame at any time. Lameness affects production as well as welfare: Archer et al. (2010) found that lame cows produced about 1 kg less milk per day. There is a wide range of clinical causes of lameness in dairy cows (Bran et al. 2019) and the clinical condition can be treated using appropriate analgesia (Stilwell et al. 2019).

The clear measurable welfare outcome of lameness is that the cow shows difficulty in walking and consequent alteration of gait. This can be identified by direct observation of the animal while walking or while attempting to move or walk. Those assessing lameness categorise walking ability using a scale ranging from no problem to scarcely able to walk for each individual. Cows that are lame walk less often to particular resources and walk less in total. Some automated measurement methods can be used, especially for the total amount of walking by the individual, but direct observation by a skilled observer is still the best measure. A criterion for lameness that is generally usable by inspectors is necessary as farmers often fail to identify lameness, especially milder cases. The extent of lameness in a herd can be assessed by an inspection visit to a farm, or to a slaughterhouse, during which every animal (or a representative sample) is checked.

2.2 Mastitis

Mastitis in mammals is a very painful condition. The teat and udder are very sensitive to touch and there is obvious diminution of normal function. Despite many advances in understanding and treatment, mastitis remains the most economically significant bacterial disease of dairy cattle (Moroni et al. 2017b). The incidence of clinical mastitis has been estimated to range from 14 to 40 cases/100 cows per year, and is worst in poorly managed herds and some high-yielding herds (EFSA 2009; Jamali et al. 2018). It has been argued that, as there have been major advances in veterinary science, the prevalence of this serious animal welfare problem might have declined more if there had been less pressure to keep increasing milk yield per cow. Information about mastitis, including sub-clinical mastitis, can be obtained by clinical examination of the cow and by other measures such as white cell count in the milk (Sant' Anna and Paranhos da Costa 2011; Moroni et al. 2017a). Herd records should be kept and be made available for evaluation as part of welfare assessment procedures. A threshold for identification of mastitis is required in order that current prevalence during an inspection visit or prevalence as assessed from herd records can be calculated. Every milking cow on a farm or other unit should be taken into account.

2.3 Reproductive disorders

Reproductive disorders are important indicators of poor welfare in dairy cows. Esslemont and Kossaibati (1997) reported that of cows that were culled, 36–44% of cows were culled because of failure to get in calf. This problem has continued to be important, especially in high-yielding cows that are metabolically stressed (Dobson et al. 2007; Berglund 2008; Friggens et al. 2010; Leroy et al. 2015; Diskin et al. 2016; Evans and Zeng 2017). Reproductive disorders, lameness and mastitis result in cows being culled earlier in life and hence having shorter lives than they did 20–30 years ago (Compton et al. 2017). Cloning of cattle has high risks of poor welfare (Broom 2014, 2018a), and so many countries do not permit it. The data required for assessing the extent of reproductive disorders on a dairy farm are those from clinical investigation during an inspection visit and those from records of veterinary treatment and from the culling of cows (von Keyserlingk et al. 2009).

2.4 Other cow welfare problems

One aspect of more intensive dairy herd management that has attracted concern is housing. Many modern housing systems for cattle involve keeping animals indoors, in a more restricted space, sometimes at high stocking density. This can increase the chance of injury, spread of disease and heat stress as well as cause poor welfare by restricting behaviour (EFSA 2009; Polsky and von Keyserlingk 2017; Rushen 2017). While cows prefer an adequate to an inadequate food supply (which can be better controlled in intensive systems), they also show clear preferences for access to pasture rather than continuous housing (Spörndly and Wredle 2004; Legrand et al. 2009; Charlton et al. 2011, 2013; Von Keyserlingk et al. 2017). Recent research has confirmed the benefits of outdoor access to pasture in allowing the expression of grazing, lying, standing, walking and social behaviours (Smid et al. 2020).

Husbandry practices that have caused concern include tail-docking (Broom and Fraser 2015), which most dairy cattle welfare programmes around the world have banned (e.g. FARM Programme in the United States), the need for caesarean section because of breed or impregnation by a bull that is large in relation to the size of the cow, and embryo loss when cloning is used. Dystocia and other abdominal pain cause poor welfare and the use of analgesia during all parturition can have beneficial effects (Stilwell et al. 2014). Hot conditions cause poor welfare in cattle and measures of heat stress in cows are described by Polsky and von Keyserlingk (2017).

2.5 Calf welfare problems

The process of parturition can be a stressful, traumatic and hazardous event for both cow and calf. It has been estimated that up to 10% of heifers, when stillbirths are included, may die before weaning in the United States with lower rates in Europe and elsewhere (Miller-Cushon et al. 2017). In the first few days after birth, the major calf welfare problems are enteric and respiratory diseases, identifiable by clinical signs. The calves of dairy cows may fail to obtain sufficient colostrum for a variety of reasons, and will then be more susceptible to disease, so checks on suckling or measurement of immunoglobulin concentration in the blood give information about welfare. The diet of the calf should have sufficient iron to prevent anaemia and, a little later in life, sufficient fibre to promote normal gut development (Phillips 2002; Broom and Fraser 2015). Both of these effects and immune system function can be used as welfare indicators.

Most dairy calves are separated from their mothers at an early age to be reared individually with potentially negative welfare consequences (Phillips 2017). In situations of inadequate management, signs of poor welfare in calves reared in individual crates or pens may include stereotypies; difficulties in standing, lying and grooming; excessive grooming of the front of the body with the ingestion of much hair and the formation of hairballs in the gut; and substantial adverse reactions to walking and to transport (Broom and Leaver 1978; Veissier et al. 1994; Broom 1996; Boe and Faerevik 2003; Gaillard et al. 2014; Broom and Fraser 2015). In contrast, group housing, if well managed, improves feeding, health, and development of behaviour and cognition

(Miller-Cushon et al. 2017). Calves may also be subjected to practices that cause pain including castration and horn-disbudding, resulting in measurable pain-related behaviours preventable by anaesthesia and analgesia (Stafford and Mellor 2005; Stilwell et al. 2008a,b, 2009, 2010, 2012).

An ethical problem for the dairy industry is the production of unwanted male calves. If these calves are removed from the cow and humanely killed, there is no welfare problem for the calf. However, if they are handled roughly, transported in inadequate ways or kept in conditions that do not provide for their needs, their welfare will be poor. These last points apply also to female calves. The solution to this problem may be the sexing of semen so that no unwanted calves are produced.

3 Dairy production, welfare and sustainability

As previously discussed, production methods can have negative welfare impacts which need to be properly understood and addressed. Different methods also have broader environmental impacts which themselves have welfare implications, such as the hotter and drier conditions resulting from climate change that livestock already encounter (Pryce and de Haas 2017). Consumers concerned about the effects of dairy products on their health can now access good information about how to manage cholesterol levels, avoid obesity and also take advantage of the nutritional benefits of dairy products. With well-controlled labelling and traceability, consumers can choose to buy products that are high welfare, have low greenhouse gas emission, are indicated as 'fair trade' and are otherwise sustainable. These issues have major consequences for the future survival and success of dairy farming.

It is well known that herbivorous animals such as cattle and sheep can consume material that humans cannot consume (e.g. grass) and convert it to an edible form (e.g. meat or milk). Many parts of the world are not suitable for arable crop production but can support pasture, suggesting that extensive, pasture-based dairy production, if properly managed, can be an efficient and sustainable use of available world resources (Marshall and Collins 2018).

However, while grasses continue to dominate feed resources for livestock production globally, feed from sources such as grain may represent around 30–40% of the diet of dairy cattle in some regions such as North America characterised by more intensive forms of production (Blümmel et al. 2017). The use of grain and other crops as feed has become controversial because of competition for land and other resources with crop production for food (Herrero et al. 2010; Broom 2018b; Balmford et al. 2018). Stoll-Kleemann and O'Riordan (2015) estimate that 45% of greenhouse gas production from cattle production comes from feed production and processing, with 39% from cow digestion and 10% from their faeces. Feeding energy-dense, highly fermentable diets to optimise milk production can also have welfare implications. If excess grain is fed, this can lead to digestive disorders such as rumen acidosis which has been estimated to have a prevalence rate of up to 60%, depending on the stage of lactation (Penner 2017). More broadly, intensive production systems have been linked to other problems such as high energy and input use, lower levels of biodiversity as well as increased rural poverty and inequality because of the consolidation of many smaller farms into a few, large, resource-intensive farming operations (Scott and Gooch 2017).

There is now a huge debate about how to make dairy farming more sustainable, both in terms of welfare and its broader environmental impact. Housed cattle fed on highly digestible crop-derived feeds have high growth rates and feed-use efficiencies. The land required for these intensive systems, taking account of land used for feed production, uses a similar amount of land to fertilised pasture systems and substantially less than extensive grazing systems (Broom 2019; Woolf 2020). High dairy production per unit of land makes more land available for other uses such as nature reserves (Balmford et al. 2018). Some research is concentrating on how to optimise the environmental benefits of this efficiency while reducing negative environmental impacts and dealing with welfare issues linked to more intensive production.

Other research has focussed on how to optimise extensive systems to increase grazing efficiency and optimise product quality while reducing greenhouse gases such as methane, protecting biodiversity and maintaining good animal welfare, including animal health (Woodfield and Judson 2018). Studies show how to change ruminant diets so as to reduce the rumen population of the microorganisms that produce methane. There is also increasing interest in the potential advantages of integrated crop-livestock systems in combatting climate change and improving resilience in agricultural production (van Wijk 2020). This includes agroforestry practices such as silvopastoralism (Moreno and Rolo 2019). Areas have been shown to produce more edible plant material and more meat or milk when trees and shrubs with edible, high protein leaves as well as pasture plants are grown in semi-intensive silvopastoral systems than pasture-only systems (Murgueitio et al. 2008; Broom et al. 2013; Broom 2017b). In some parts of the world where climate would be favourable, dairy cattle could, for example, be managed on pastures with shrubs such as *Leucaena leucocephala* (Radrizzani et al. 2016). Cut leafy branches of trees such as ramón *Brosimum alicastrum* can be an important source of food where there is a dry season (Ku-Vera et al. 2013). Other benefits are that biodiversity is much higher in semi-intensive silvopastoral systems than in pasture-only systems, with more tick and fly predators so reducing cattle disease, as well as the potential to better conserve forest resources (Burgess 1999; Múnera et al. 2008; Rivera et al. 2008; McAdam and McEvoy 2009; Sutherland et al. 2018).

4 Conclusion

This chapter has highlighted some of the key welfare issues faced by the dairy sector. It has placed these challenges within the broader framework of the sustainability of the sector which is under growing scrutiny. It is important to emphasise that, while there are some serious welfare problems at present, in several systems dairy cows and calves can be produced with very good welfare. Retailers and all who buy dairy products should insist on seeing evidence for high welfare standards. Whether it is welfare or environmental impact, transparency is essential. Consumers need to obtain accurate information. The best dairy production systems use resources that cannot easily be used in any other way for food production, have no negative impacts on animal welfare and have no greater negative impact on the environment than wholly plant-based alternatives. Much change in the industry is needed in order to find and use the best production methods to achieve these challenging goals.

5 Where to look for further information

A comprehensive review of dairy cow welfare by a group of independent scientists is that by the European Food Safety Authority (EFSA) (2009). Some other aspects of sustainability are reviewed by De Jong (2013) and Balmford et al. (2018).

6 References

Archer, S. C., Green, M. J. and Huxley, J. N. (2010). Association between milk yield and serial locomotion score assessments in UK dairy cows. *J. Dairy Sci.* 93(9), 4045–4053.

Balmford, A., Amano, T., Bartlett, H., Chadwick, D., Collins, A., Edwards, D., Field, R., Garnsworthy, P., Green, R., Smith, P., Waters, H., Whitmore, A., Broom, D. M., Chara, J., Finch, T., Garnett, E., Gathorne-Hardy, A., Hernandez-Medrano, J., Herrero, M., Hua, F., Latawiec, A., Misselbrook, T., Phalan, B., Simmons, B. I., Takahashi, T., Vause, J., Zu Ermgassen, E. and Eisner, R. (2018). The environmental costs and benefits of high-yield farming. *Nat. Sustainability* 1(9), 477–485. doi:10.1038/s41893-018-0138-5.

Barker, Z. E., Leach, K. A., Whay, H. R., Bell, N. J. and Main, D. C. J. (2010). Assessment of lameness prevalence and associated risk factors in dairy herds in England and Wales. *J. Dairy Sci.* 93(3), 932–941.

Bell, N. (2017). Preventing and managing lameness in dairy cows. In: Webster, J. (Ed.), *Achieving Sustainable Production of Milk Volume 3: Dairy Herd Management and Welfare*. Burleigh Dodds Science Publishing, Cambridge, UK.

Bennett, R. M., Anderson, J. and Blaney, R. J. P. (2002). Moral intensity and willingness to pay concerning farm animal welfare issues and the implications for agricultural policy. *J. Agric. Environ. Eth.* 15(2), 187–202.

Berglund, B. (2008). Genetic improvement of dairy cow reproductive performance. *Reprod. Domest. Anim.* 43 (Suppl. 2), 89–95. Available at: doi.org/10.1111/j.1439-0531.2008.01147.x.

Blümmel, M., Muller, A., Schader, C., Herrero, M. T. and Garg, M. R. (2017). The use and abuse of cereals, legumes and crop residues in rations for dairy cattle. In: Webster, J. (Ed.), *Achieving Sustainable Production of Milk Volume 3: Dairy Herd Management and Welfare*. Burleigh Dodds Science Publishing, Cambridge, UK.

Boe, K. E. and Faerevik, G. (2003). Grouping and social preferences in calves, heifers and cows. *Appl. Anim. Behav. Sci.* 80(3), 175–190.

Bran, J. A., Costa, J. H. C., von Keyserlingk, M. A. G. and Hötzel, M. J. (2019). Factors associated with lameness prevalence in lactating cows housed in freestall and compost-bedded pack dairy farms in southern Brazil. *Prev. Vet. Med.* 172, 104773, doi:10.1016/j.prevetmed.2019.104773.

Broom, D. M. (1996). Scientific research on veal calf welfare. In: *Veal Perspectives to the Year 2000*. Proceedings of the International Symposium, Le Mans, France. Fédération de la Vitellerie Francaise, Paris, pp. 147–153.

Broom, D. M. (2001). The use of the concept animal welfare in European conventions, regulations and directives. In: *Food Chain*. SLU Services, Uppsala, pp. 148–151.

Broom, D. M. (2010). Animal welfare: an aspect of care, sustainability, and food quality required by the public. *J. Vet. Med. Educ.* 37(1), 83–88. doi:10.3138/jvme.37.1.83.

Broom, D. M. (2013). Cow welfare and herd size: towards a sustainable dairy industry. *Cattle Pract.* 21, 169–173.

Broom, D. M. (2014). *Sentience and Animal Welfare*. CABI Publishing, Wallingford, p. 200.

Broom, D. M. (2017a). *Animal Welfare in the European Union*. European Parliament Policy Department, Brussels, Belgium. p. 75. ISBN: 978-92-846-0543-9.

Broom, D. M. (2017b). Components of sustainable animal production and the use of silvopastoral systems. *R. Bras. Zootec.* 46(8), 683–688, doi:10.1590/s1806-92902017000800009.

Broom, D. M. (2018a). Animal welfare and the brave new world of modifying animals. In: Grandin, T. and Whiting, M. (Eds), *Are We Pushing Animals to Their Biological Limits?* CABI Publishing, Wallingford, pp. 172–180.

Broom, D. M. (2018b). The scientific basis for action on animal welfare and other aspects of sustainability. In: D'Silva, J. and McKenna, C. (Eds), *Farming, Food and Nature: Respecting Animals, People and the Environment*. Earthscan Publications, London, UK; Routledge, New York, NY, USA, pp. 93–100, ISBN: 9781138541443.

Broom, D. M. (2019). Land and water usage in beef production systems. *Animals* 9(6), 286. doi:10.3390/ani9060286.

Broom, D. M. and Fraser, A. F. (2015). *Domestic Animal Behaviour and Welfare* (5th edn). CABI Publishing, Wallingford, p. 472.

Broom, D. M., Galindo, F. A. and Murgueitio, E. (2013). Sustainable, efficient livestock production with high biodiversity and good welfare for animals. *Proc. Biol. Sci.* 280(1771), 20132025. doi:10.1098/rspb.2013.2025.

Broom, D. M. and Leaver, J. D. (1978). The effects of group-housing or partial isolation on later social behaviour of calves. *Anim. Behav.* 26, 1255–1263.

Burgess, P. J. (1999). Effects of agroforestry on farm biodiversity in the UK. *Scot. For.* 53, 24–27.

Charlton, G. L., Rutter, S. M., East, M. and Sinclair, L. A. (2011). Effects of providing total mixed rations indoors and on pasture of the behavior of lactating dairy cattle and their preference to be indoors or on pasture. *J. Dairy Sci.* 94(8), 3875–3884.

Charlton, G. L., Rutter, S. M., East, M. and Sinclair, L. A. (2013). The motivation of dairy cows for access to pasture. *J. Dairy Sci.* 96(7), 4387–4396.

Compton, C. W. R., Heuer, C., Thomsen, P. T., Carpenter, T. E., Phyn, C. V. C. and McDougall, S. (2017). Invited review: a systematic literature review and meta-analysis of mortality and culling in dairy cattle. *J. Dairy Sci.* 100(1), 1–16.

Cook, N. B., Hess, J. P., Foy, M. R., Bennett, T. B. and Brotzman, R. L. (2016). Management characteristics, lameness, and body injuries of dairy cattle housed in high-performance dairy herds in Wisconsin. *J. Dairy Sci.* 99(7), 5879–5891.

De Jong, P. (Ed.) (2013). *Sustainable Dairy Production*. John Wiley & Sons, Chichester.

Diskin, M. G., Waters, S. M., Parr, M. H. and Kenny, D. A. (2016). Pregnancy losses in cattle: potential for improvement. *Reprod. Fertil. Dev.* 28(1–2), 83–93.

Dobson, H., Smith, R. F., Royal, M. D., Knight, Ch. and Sheldon, I. M. (2007). The high-producing dairy cow and its reproductive performance. *Reprod. Domest. Anim.* 42(Suppl. 2), 17–23. doi:10.1111/j.1439-0531.2007.00906.x.

EFSA (2009). Scientific report on the overall effects of farming systems on dairy cow welfare and disease. *Annex to the EFSA J.* 1143, 1–38.

Esslemont, R. J. and Kossaibati, M. A. (1997). Culling in 50 dairy herds in England. *Vet. Rec.* 140, 36–39.

EU SCAHAW (1999). *EU Scientific Committee on Animal Health and Animal Welfare Report on Animal Welfare Aspects of the Use of Bovine Somatotrophin*. European Commission, Brussels.

Evans, A. and Zeng, S. (2017). Causes, prevention and management of infertility in dairy cows. In: Webster, J. (Ed.), *Achieving Sustainable Production of Milk Volume 3: Dairy Herd Management and Welfare*. Burleigh Dodds Science Publishing, Cambridge, UK.

Friggens, N. C., Disenhaus, C. and Petit, H. V. (2010). Nutritional sub-fertility in the dairy cow: towards improved reproductive management through a better biological understanding. *Animal* 4(7), 1197–1213. doi:10.1017/S1751731109991601.

Gaillard, C., Meagher, R. K., von Keyserlingk, M. A. and Weary, D. M. (2014). Social housing improves dairy calves' performance in two cognitive tests. *PLoS ONE* 9(2), e90205 (6 pages).

Herrero, M., Thornton, P. K., Notenbaert, A. M., Wood, S., Msangi, S., Freeman, H. A., Bossio, D., Dixon, J., Peters, M., van de Steeg, J., Lynam, J., Parthasarathy Rao, P., Macmillan, S., Gerard, B., McDermott, J., Seré, C. and Rosegrant, M. (2010). Smart investments in sustainable food production: revisiting mixed crop-livestock systems. *Science* 327(5967), 822–825, doi:10.1126/science.1183725.

Jamali, H., Barkema, H. W., Jacques, M., Lavallée-Bourget, E. M., Malouin, F., Saini, V., Stryhn, H. and Dufour, S. (2018). Invited review: incidence, risk factors and effect of clinical mastitis recurrence in dairy cows. *J. Dairy Sci.* 101(6), 4729–4746.

Knaus, W. (2009). Dairy cows trapped between performance demands and adaptability. *J. Sci. Food Agric.* 89(7), 1107–1114.

Ku-Vera, J. C., Ayala-Burgos, A. J., Solorio-Sánchez, F. J., Briceño-Poot, E. G., Ruiz-González, A., Piñeiro-Vázquez, A. T., Barros-Rodríguez, M., Soto-Aguilar, A., Espinoza-Hernández, J. C., Albores-Moreno, S. and Chay-Canul, A. J. (2013). Tropical tree foliages and shrubs as feed additives in ruminant rations. In: *Nutritional Strategies of Animal Feed Additives,ed. A.F.Z.M. Salem*. Nova Science Publishers, New York, USA, pp. 59-76.

Legrand, A. L., von Keyserlingk, M. A. and Weary, D. M. (2009). Preference and usage of pasture versus free-stall housing by lactating dairy cattle. *J. Dairy Sci.* 92(8), 3651-3658.

Leroy, J. L., Valckx, S. D., Jordaens, L., De Bie, J., Desmet, K. L., Van Hoeck, V., Britt, J. H., Marei, W. F. and Bols, P. E. (2015). Nutrition and maternal metabolic health in relation to oocyte and embryo quality: critical views on what we learned from the dairy cow model. *Reprod. Fertil. Dev.* 27(4), 693-703. doi:10.1071/RD14363.

Marshall, A. and Collins, R. (Eds) (2018). *Improving Grassland and Pasture Management in Temperate Agriculture*. Burleigh Dodds Science Publishing, Cambridge, UK.

McAdam, J. H. and McEvoy, P. M. (2009). The potential for silvopastoralism to enhance biodiversity on grassland farms in Ireland. In: Rigueiro-Rodriguez, A., McAdam, J. and Mosquera-Losada, M. R. (Eds), *Agroforestry in Europe: Current Status and Future Prospects*. Springer, Berlin, Germany, pp. 343-356.

Miller-Cushon, E., Leslie, K. and DeVries, T. (2017. Ensuring the health and welfare of dairy calves and heifers. In: Webster, J. (Ed.), *Achieving Sustainable Production of Milk Volume 3: Dairy Herd Management and Welfare*. Burleigh Dodds Science Publishing, Cambridge, UK, p. 149.

Moreno, G. and Rolo, V. (2019). Agroforestry practices: silvopastoralism. In: Mosquera-Losada, M. and Prabhu, R. (Eds), *Agroforestry for Sustainable Agriculture*. Burleigh Dodds Science Publishing, Cambridge, UK.

Moroni, P. (2017a). Aetiology, diagnosis and control of mastitis in dairy herds. In: Webster, J. (Ed.), *Achieving Sustainable Production of Milk Volume 3: Dairy Herd Management and Welfare*. Burleigh Dodds Science Publishing, Cambridge, UK.

Moroni, P. (2017b). Mastitis, milk yield and quality. In: In van Belzen, N. (Ed.), *Achieving Sustainable Production of Milk Volume 1: Safety, Quality and Sustainability*. Burleigh Dodds Science Publishing, Cambridge, UK.

Múnera, E., Bock, B. C., Bolivar, D. M. and Botero, J. A. (2008). Composición y estructura de la avifauna en diferentes hábitats en el departamento de Córdoba, Colombia. In: Murgueitio, E., Cuartas, C. A. and Naranjo, J. F. (Eds), *Ganadería del Futuro*. Fundacio´n CIPAV, Cali, Colombia, pp. 205-225.

Murgueitio, E., Cuartas, C. A. and Naranjo, J. F. (2008). *Ganadería del Futuro*. Fundación CIPAV, Cali Colombia.

Oltenacu, P. A. and Broom, D. M. (2010). The impact of genetic selection for increased milk yield on the welfare of dairy cows. *Anim. Welf.* 19(S), 39-49.

Penner, G. (2017). Disorders of digestion and metabolism in dairy cattle: the case of subacute rumen acidosis. In: Webster, J. (Ed.), *Achieving Sustainable Production of Milk Volume 3: Dairy Herd Management and Welfare*. Burleigh Dodds Science Publishing, Cambridge, UK.

Phillips, C. J. C. (2002). *Cattle Behaviour and Welfare* (2nd edn.). Blackwell Scientific Publications, Oxford, U.K.

Phillips, C. (2017). Understanding the behaviour of dairy cattle. In: Webster, J. (Ed.), *Achieving Sustainable Production of Milk Volume 3: Dairy Herd Management and Welfare*. Burleigh Dodds Science Publishing, Cambridge, UK.

Polsky, L. and von Keyserlingk, M. A. G. (2017). Invited review: effects of heat stress on dairy cattle welfare. *J. Dairy Sci.* 100(11), 8645-8657. doi:10.3168/jds.2017-12651.

Pryce, J. and de Haas, Y. (2017). Genetic selection for dairy cow welfare and resilience to climate change. In: Webster, J. (Ed.), *Achieving Sustainable Production of Milk*

Volume 3: Dairy Herd Management and Welfare. Burleigh Dodds Science Publishing, Cambridge, UK.

Radrizzani, A., Shelton, H. M., Kravchuk, O. and Dalzell, S. A. (2016). Survey of long-term productivity and nutritional status of *Leucaena leucocephala*-grass pastures in subtropical Queensland. *Anim. Prod. Sci.* 56(12), 2064–2073. doi:10.1071/AN15084.

Ranjbar, S., Rabiee, A. R., Gunn, A. and House, J. K. (2016). Identifying risk factors associated with lameness in pasture-fed dairy herds. *J. Dairy Sci.* 99(9), 7495–7505.

Rivera, L. B., Botero, M., Escobar, S. and Armbrecht, I. (2008). Diversidad de hormigas en sistemas ganaderos. In: Murgueitio, E., Cuartas, C. A. and Naranjo, J. F. (Eds), *Ganadería del Futuro*. Fundación CIPAV, Cali, Colombia, pp. 228–243.

Robbins, J. A., von Keyserlingk, M. A. G., Fraser, D. and Weary, D. M. (2016). Invited review: farm size and animal welfare. *J. Anim. Sci.* 94(12), 5439–5455.

Rushen, J. (2017). Housing and the welfare of dairy cattle. In: Webster, J. (Ed.), *Achieving Sustainable Production of Milk Volume 3: Dairy Herd Management and Welfare*. Burleigh Dodds Science Publishing, Cambridge, UK.

Sant'Anna, A. C. and Paranhos da Costa, M. J. R. (2011). The relationship between dairy cow hygiene and somatic cell count in milk. *J. Dairy Sci.* 94(8), 3835–3844, doi:10.3168/jds.2010-3951.

Schukken, Y. H., Tikofsky, L. L. and Zadoks, R. N. (2005). Environmental control for mastitis prevention, milk quality and food safety. In: Hogeveen, H. (Ed.), *Mastitis in Dairy Production: Current Knowledge and Future Solutions*. Wageningen Academic Publishers, Wageningen, The Netherlands, pp. 109–114.

Scott, N. and Gooch, C. (2017). 'Towards' sustainability of dairy farming: an over. In: Invan Belzen, N. (Ed.), *Achieving Sustainable Production of Milk Volume 2: Safety, Quality and Sustainability*. Burleigh Dodds Science Publishing, Cambridge, UK.

Smid, A. C., Weary, D. M. and von Keyserlingk, M. A. G. (2020). The influence of different types of outdoor access on dairy cattle behaviour. *Front. Vet. Sci.* 13 May 2020. Available at: https//:doi.org/10.3309/fvets.2020.00257.

Spörndly, E. and Wredle, E. (2004). Automatic milking and grazing – effects of distance to pasture and level of supplements on milk yield and cow behavior. *J. Dairy Sci.* 87(6), 1702–1712.

Stafford, K. J. and Mellor, D. J. (2005). Dehorning and disbudding distress and its alleviation in calves. *Vet. J.* 169(3), 337–349.

Stilwell, G., Carvalho, R. C., Carolino, N., Lima, M. S. and Broom, D. M. (2010). Effect of hot-iron disbudding on behaviour and plasma cortisol of calves sedated with xylazine. *Res. Vet. Sci.* 88(1), 188–193.

Stilwell, G., Carvalho, R. C., Lima, M. S. and Broom, D. M. (2009). Effect of caustic paste disbudding, using local anaesthesia with and without analgesia, on behaviour and cortisol of calves. *Appl. Anim. Behav. Sci.* 116(1), 35–44.

Stilwell, G. T., Ferrador, A. M., Santos, M. S., Domingues, J. M. and Carolino, N. (2019). Use of topical local anesthetics to control pain during treatment of hoof lesions in dairy cows. *J. Dairy Sci.* 102(7), 6383–6390.

Stilwell, G., Lima, M. S. and Broom, D. M. (2008a). Comparing plasma cortisol and behaviour of calves dehorned with caustic paste after non-steroidal-anti-inflammatory analgesia. *Livest. Sci.* 119(1–3), 63–69.

Stilwell, G., Lima, M. S. and Broom, D. M. (2008b). Effects of nonsteroidal anti-inflammatory drugs on long-term pain in calves castrated by use of an external clamping technique following epidural anaesthesia. *Am. J. Vet. Res.* 69(6), 744–750.

Stilwell, G., Lima, M. S., Carvalho, R. C. and Broom, D. M. (2012). Effects of hot-iron disbudding, using regional anaesthesia with and without carprofen, on cortisol and behaviour of calves. *Res. Vet. Sci.* 92(2), 338–341.

Stilwell, G., Schubert, H. and Broom, D. M. (2014). Effects of analgesic use post-calving on cow welfare and production. *J. Dairy Sci.* 97(2), 888–891.

Stoll-Kleemann, S. and O'Riordan, T. (2015). The sustainability challenges of our meat and dairy diets. *Environ. Sci. Policy Sustain. Dev.* 57(3), 34–48.

Sutherland, W. J., Dicks, L. V., Everard, M. and Geneletti, D. (2018). Qualitative methods for ecologists and conservation scientists. *Methods Ecol. Evol.* 9(1), 7–9.

Van der Werf, J. and Pryce, J. (Eds) (2019). *Advances in Breeding of Dairy Cattle*. Burleigh Dodds Science Publishing, Cambridge, UK.

Van Wijk, M. (2020). The contribution of integrated crop-livestock systems in combatting climate change and improving resilience in agricultural production. In: Deryng, D. (Ed.), *Climate Change and Agriculture*. Burleigh Dodds Science Publishing, Cambridge, UK.

Veissier, I., Gesmier, V., Le Neindre, P., Gautier, J. Y. and Bertrand, G. (1994). The effects of rearing in individual crates on subsequent social behaviour of veal calves. *Appl. Anim. Behav. Sci.* 41(3–4), 199–210.

Von Keyserlingk, M. A., Cestari, A. A., Franks, B., Fregonesi, J. A. and Weary, D. M. (2017). Dairy cows value access to pasture as highly as fresh feed. *Sci. Rep.* 7, 44953.

Von Keyserlingk, M. A. G., Martin, N. P., Kebreab, E., Knowlton, K. F., Grant, R. J., Stephenson, M., Sniffen, C. J., Harner, J. P., Wright, A. D. and Smith, S. I. (2013). Invited review: sustainability of the US dairy industry. *J. Dairy Sci.* 96(9), 5405–5425.

Von Keyserlingk, M. A. G., Rushen, J., de Passillé, A. M. and Weary, D. M. (2009). Invited review: the welfare of dairy cattle–key concepts and the role of science. *J. Dairy Sci.* 92(9), 4101–4111. doi:10.3168/jds.2009-2326.

Webster, J. (2017). *Achieving Sustainable Production of Milk. Volume 3: Dairy Herd Management and Welfare*. Burleigh Dodds Science Publishing Limited, Cambridge, UK.

Woodfield, D. and Judson, H. (2018). Balancing pasture productivity with environmental and animal health requirements. In: Marshall, A. and Collins, R. (Eds), *Improving Grassland and Pasture Management in Temperate Agriculture*. Burleigh Dodds Science Publishing, Cambridge, UK.

Woolf, J. (2020). Quantifying the role of livestock in climate change. In: Deryng, D. (Ed.), *Climate Change and Agriculture*. Burleigh Dodds Science Publishing, Cambridge, UK.

Part 1

Understanding behaviour

Chapter 2

Advances in understanding cognition and learning in cattle

Maria Vilain Rørvang, Swedish University of Agricultural Sciences, Sweden; and Christian Nawroth, Leibniz Institute for Farm Animal Biology, Germany

1 Introduction

Recent animal welfare concepts changed from more traditional frameworks such as the five freedoms (Farm Animal Welfare Council, 2013) to also include animals' needs (Špinka, 2006), affective states (Mendl et al., 2010) and inter-individual behavioral differences (Finkemeier et al., 2018). Collectively, these concepts emphasize the need for detailed knowledge about how farm animals perceive and interact with their environment.

Cattle perceive their surroundings via their senses, that is, olfaction, hearing, vision, taste and touch. All signals are processed by the brain to produce information that the cow can act upon and store. Cognition then refers to the processes used by the cow to generate adaptive and flexible responses when reacting to the surroundings (Shettleworth, 2010). Such responses can be both innate and learned (instinct vs. experience, respectively).

Cattle may, as other non-human animals, differ from humans in how they perceive and learn from their environment. As a result, we as humans may not always consider the cow's perspective when designing facilities for cattle or when handling cattle. Knowledge of this area, and the practical implications this may have, is therefore essential to ensure adapted management and handling

http://dx.doi.org/10.19103/AS.2020.0084.02

practices, safeguarding the work environment (for both animals and handlers), and to sustain production, health and welfare.

2 Learning and memory in cattle

The process of learning arises from the motivation to acquire information through experience. Many of the behaviors performed by conventional cattle in their daily life are acquired by learning processes, such as managing how to be milked by a milking robot. These behaviors can be learned in several ways, for example solely by trial and error, directly from the stock person or from fellow herd mates.

2.1 Feature cue learning and memory

Cows can discriminate a wide range of stimuli and have good capacities for visual discrimination learning: in operant conditioning discrimination paradigms, they can distinguish between geometric shapes (Baldwin, 1981), colors (Gilbert and Arave, 1986), size (Rehkämper and Görlach, 1997) and stimuli differing by both brightness and size (Schaeffer and Sikes, 1971). They also remember the association between a visual feature cue and a reward for up to one year (Hirata and Takeno, 2014) (Fig. 1).

Figure 1 Cows are highly trainable and can, for example, be trained to follow a target like the black bucket in the picture. Even so, classical training principles such as positive reinforcement for training cattle have not been systematically used and hence there may be unexploited potential to use these well-developed training regimes. Photo: Maria Vilain Rørvang.

Cattle can flexibly use the learned associations with visual cues in several foraging contexts (Howery et al., 2000), including the transfer of this knowledge from the pen into the field (Renken et al., 2008). However, these findings need a cautious interpretation: experiments using more naturalistic stimuli (e.g. vegetation patches) indicate that the ability of cows to visually discriminate vegetation patches when foraging in grasslands is likely not as high as that expected from the visual acuity reported in previous studies (Hirata et al., 2019).

2.2 Spatial learning and memory

Cows as grazing ungulates should possess good spatial memory, which is used, for example, when finding and relocating food patches to increase feeding efficiency. It is thus not surprising that dairy cows have been found to locate desirable food resources from a distance (Hirata et al., 2019) and use spatial memory to relocate and remember the location of specific food resources (Bailey and Sims, 1998; Laca, 1998).

Cattle also remember the location of food for up to 40 days (Ksiksi and Laca, 2002) and can track several locations with food resources, and even remember these locations for up to 8 h (Bailey et al., 1989). Heifers and cows did not only quickly learn which side of a modified T-maze contained grain (Kovalčik and Kovalčik, 1986); some cows also quickly mastered more complex mazes and retained the memory of the maze configuration for up to 6 weeks (Hirata et al., 2016). Cows are able to associate locations with the quantity and quality of the particular food found (Bailey and Sims, 1998), flexibly adjusting their foraging patterns to take advantage of this knowledge.

3 Understanding the physical environment

The term 'physical cognition' refers to an organism's understanding of objects and their various spatial and causal relationships. Currently, not much is known about how cattle understand their physical environment. There are indications of cattle being able to discriminate between not only the type but also the amount of food located at a food patch (Bailey and Sims, 1998), showing that they can differentiate between different quantities. Cattle also form social categories, such as 'familiar' and 'unfamiliar' individuals (Coulon et al., 2011), demonstrating that they have the ability to categorize. At present, it is unknown if cattle possess other sophisticated physico-cognitive capacities, but studies on object permanence in horses and goats (Uller and Lewis, 2009; Nawroth et al., 2015) and on number discrimination in horses (Gabor and Gerken, 2014) hint these capacities to be widespread in other ungulate farm species.

4 Social cognition in the cattle group

Like physical objects, cows also have to (re)locate and identify conspecifics. As social entities, however, other subjects also create a need for additional decision-making processes, which are not present in the world of inanimate objects. For example, cows have to form and establish stable social groups, while other individuals could also be a source of new information to learn. In addition, other cows can behave spontaneously, and hence developing mechanisms that enable the cow to anticipate future behavior of herd mates is key.

4.1 Social recognition

Cattle possess the ability to discriminate between and recognize each other. Such capacities are evident, for example, during regrouping at abattoirs, where heifers express less aggressive behavior toward familiar individuals compared to unfamiliar individuals (Bouissou and Andrieu, 1978). Cattle use visual, olfactory and auditory cues to discriminate individuals (Hagen and Broom, 2003) and can be trained to perform the discrimination based solely on olfactory cues (Baldwin, 1977). Coulon et al. (2007) showed that cattle can differentiate their own species from other species and later Coulon et al. (2010) showed that cattle are able to discriminate between kin and non-kin individuals using the visual morphological characteristics of each individual. Another study on this topic has shown that morphological characteristics of the head of a conspecific are sufficient for social discrimination in cattle, in this case presented as 2D images (Coulon et al., 2011). In addition, subordinate cattle often avoid more dominant individuals, suggesting that they have formed an association with a particular individual based on its previous experience (Bouissou and Andrieu, 1977). Bouissou and Andrieu (1978) further noted that cattle, who were previously observed within close proximity of each other, or allogrooming, were also commonly the preferred social partners. Collectively, these abilities are of evolutionary importance for the cohesion of the cattle group and for maintaining a low level of aggression. In practical terms, keeping the cattle group (e.g. milking herd, grazing herd) stable is important to ensure a low level of agitation and conflict between the animals.

Another important aspect of social recognition is between the cow and her calf. Several studies have demonstrated that calves, even from an early age, can differentiate their mothers' call from other cows (Marchant-Forde et al., 2002; Barfield et al., 2010; Padilla De La Torre et al., 2016), and that mothers can distinguish their calves' calls from other calves (Padilla De La Torre et al., 2016). The latter is influenced by the age of the calf, resulting in mothers being decreasingly responsive with increasing age (Padilla De La Torre et al., 2016).

Olfactory recognition also plays a crucial role, but perhaps mainly during the first hours of the calf's life. Immediately after calving, the cow will use substantial time licking the fur of the newborn (Jensen, 2012), and failure to lick will result in reduced or even entirely absent maternal behavior and bonding, thought to be a result of the cow not learning the odor of her calf (von Keyserlingk and Weary, 2007). Odors contained in amniotic fluid in the fur of the newborn, are also thought to direct the attention of the mother to the newborn. Although the mechanism is not completely understood, this may be a learned or innate response (reviewed by Rørvang, 2018, Fig. 2.).

Figure 2 Olfactory recognition plays a crucial role in maternal recognition. Right after calving, the cow will lick and smell the fur of the newborn, which will elicit maternal behavior and bonding between cow and calf. Odors contained in amniotic fluid in the fur of the newborn, are thought to direct the attention of the mother to the newborn, which may be either a learned or an innate response. Photo: Maria Vilain Rørvang.

4.2 Learning from others

When animals observe and learn from others, it allows the acquisition of locally adaptive information from conspecifics, without risking the challenges and costs associated with individual associative learning such as a higher predation risk (Galef and Laland, 2005). Animals use a diversity of mechanisms to learn from others; these can include social facilitation, stimulus and local enhancement (collectively 'social transmission'), or goal emulation and imitation (collectively 'social learning', Heyes, 1994; Rørvang et al., 2018a). Such mechanisms have been extensively studied in several mammalian taxa, such as primates (Whiten, 2000) and canids (Pongrácz et al., 2001), but research on this topic is surprisingly sparse in ruminants.

The social organization of free-ranging cattle often comprises the inclusion of males, females and precocial young into mixed groups, with the number of males varying according to season (Bouissou et al., 2001). The relatively short social distance between individuals offers an opportunity to watch and learn from others. A limiting factor to this opportunity might, however, be the visual capacities of cattle. Like most other grazing mammals, the eyes of cattle are laterally placed on either side of the head, offering a wide-angle, panoramic view with limited binocular vision. This means that the visual acuity of cattle may be limited to the angle right in front of the cow, where stereopsis is possible. In order to observe and learn from others, the cow thus need to be in a close distance and specific position to observe the event, which inevitably lowers the chances of social transmission of information. Nonetheless, there appears to be evidence for social transmission of behavior in cattle. For example, it is commonly noted that within grazing herds, individuals synchronize their behavior, for example, when collectively moving or playing (e.g. Benham, 1982; Miller and Wood-Gush, 1991, Fig. 3). In this case, the mechanism at play is social facilitation, which relies on a behavior being contagious. In addition, Stoye et al. (2012) found that the posture of pastured, young cattle was more similar to their nearest neighbor than that of a random member of the herd, further implying that social facilitation may be most pronounced between physically close individuals – perhaps due to the visual restrictions mentioned at the beginning of this section. In terms of feed selection, Pfister et al. (2002) found that social facilitation causes cattle to start eating otherwise novel and even previously aversive food items (in this case, locoweed), and such conditioned food aversions can moreover be eliminated following exposure to non-averted social companions (Ralph and Olsen, 1990). Another example of this is when calves learn where to graze from their dam (Howery et al., 1998), which can be a result of both social facilitation and stimulus enhancement. When watching the dam graze, social facilitation will make the calf more likely to graze itself. But to a naive calf, that is, one who has not learned yet how

Figure 3 In a herd of cattle, individuals often synchronize their activities. When one cow starts grazing or playing, another is more likely to engage in the same activity. This implies that the close cohesion of the individual within the herd may offer an opportunity to observe and learn from each other. Photo: Maria Vilain Rørvang.

to graze, stimulus enhancement will make the calf more likely to approach and touch the grass, and hence more likely to learn how and what to graze by watching its dam.

Considering social learning mechanisms (goal emulation and imitation), studies on cattle are sparse, perhaps because designing such experimental tests is not a straightforward task. Veissier (1993) investigated social learning of an operant task in heifers in four consecutive experiments, and found that although observer heifers were provided with a trained demonstrator, they did not learn the task (pushing a panel to obtain food from a box) faster. On the other hand, the observer heifers did spend more time with the box and panel, implying some level of social transmission of information (in this case, via stimulus and/or local enhancement) when solving operant tasks. Such skills could, for example, be beneficial when cows learn how to operate automatic feeding bins or other automatic fixtures. By watching others maneuver the experimental apparatus, the naive cow may be more likely to approach and engage in activities that can enable it to solve the task. It is, however, also fair to argue that solving an operant task may be of less biological relevance for cattle, and hence this particular ability may not be the best way to test for social learning skills in cattle. Developing social learning tests, geared to test for social learning mechanisms, should therefore take advantage of a biologically relevant task for cattle, such as predator avoidance/detection or relocating foraging patches.

Another way to investigate social learning in cattle could be to study the transfer of information between mother and offspring as proposed by Rørvang et al. (2018b). In this situation, the cost of social learning mechanisms outweighs that of individual associative learning, as the information from the experienced dam is highly likely to be vital for the survival of the calf. If future studies are able to show that social learning mechanisms are at play when calves learn from their dam, such information would be central to the discussion of whether or not to keep cows and calves together for a longer time after calving. The calf might benefit from learning from its mother during a prolonged cow-calf contact period (Fig. 4); however, it is often standard practice in commercial dairy housing to separate the calf from its dam immediately after it is born.

4.3 Socially contagious behavior

Another aspect of socially transferable information, which could have implications for modern management of cattle, is in relation to transmittable fear reactions. Fear reactions are generally socially contagious (e.g. Griffin, 2004), but fear reactions can also be socially attenuated. Boissy and Le Neindre (1990) found that calm companions lowered the fear responses of naive heifers, and that social buffering through the mere presence of a companion allowed the heifers to develop appropriate responses to the stimuli provided.

Figure 4 One area where social learning mechanisms might be at play is when calves learn from their dam. In this situation, social learning mechanisms can be adaptive as individual associative learning has a high cost. If calves are able to utilize social learning mechanisms when learning from their dam, such information would be essential in the discussion of whether or not to keep cows and calves together. Photo: Maria Vilain Rørvang.

Future research could therefore investigate whether attenuation of fear responses can work at the group level in cattle. Such abilities hold potential to lower fearfulness in cow groups in general, but also in specific fear-eliciting situations such as hoof trimming. A complicating factor, however, is that fear reactions can function as conditioning stimuli. This means that when exposed to a fearful companion, the observing animal can associate fear with a specific situation, object or species (Olsson and Phelps, 2007). Such mechanisms have been widely studied in rodents and humans but are yet to be studied in farm animal species, and hence more research is needed to explore such mechanisms in cattle.

With regard to the control of social fear, it has been argued that visual cues are the main modality mediating attenuation (e.g. Guzmán et al., 2009), but it is possible that other mediators such as odors and vocalizations also contribute. Cattle have been found to react more fearful during feeding and exploration when exposed to olfactory cues from stressed conspecifics (Terlouw et al., 1998). Despite these findings, research on the sensory abilities of cattle is sparse, highlighting a need for more knowledge on how information is transferred between cows and how it affects individuals in the herd.

5 Human-cow interactions

Humans are an important factor in the environment of any production animal. In commercial cattle production, handlers are interacting with cows on a daily basis. Knowledge on how cows see and mentally represent humans is, therefore, central to improving handling routines and welfare.

Cattle can discriminate between humans using various visual cues (Taylor and Davis, 1998), such as a portion of their face or their body (Rybarczyk et al., 2001), or the color of clothes on unfamiliar people (Munksgaard et al., 1999). Calves can readily discriminate human handlers based on previous experience and can, after being exposed to a negative experience, develop a more generalized fear reaction to human handlers (de Passillé et al., 1996). It is thus important to establish a positive human–animal relation already at a young age, which may last throughout production, and thus have a long-term positive impact on both human and animal welfare (Rushen et al., 1999; Breuer et al., 2003). Roughly speaking, the number of positive human experiences has to outweigh that of negative ones (Hemsworth, 2003); but what motivates a cow to interact with humans in the first place? Habituation appears to be at least equally as important as positive reinforcement for the acceptance of human contact by cattle (Boivin et al., 1998). Despite its trainability, the use of learning theory as positive reinforcement training for cattle to cope in various potentially stressful situations has not been made in a systematic way as is often done, for example, in horses. There may thus be unexploited potential to use well-developed training regimes from other species (e.g. Rørvang et al., 2018c, Fig. 1).

Several studies have investigated the human–cow relationship and ways to improve it. While most of these have focused exclusively on how tactile and visual stimulation can affect this relationship, humans and cattle can also interact via other modalities such as olfaction and audition. Cattle have a more sensitive hearing ability than do humans (Heffner and Heffner, 1983, 1992) and their sense of smell plays a central role in both social and sexual behavior (Wyatt, 2003). It is therefore likely that cattle gather much more information from human handlers than is currently considered. If, for instance, a cow is able to detect the emotional expressions of a caretaker, this might affect the cow's behavior. From the management of other domestic ungulates, such as horses, there is a common saying that horses can 'smell' or 'feel' a person's bad mood. Results are conflicting, but some studies indeed show that horses handled by calm and positive humans show lower heart rates and lower cortisol concentrations (reviewed in Rørvang et al., 2020).

Based on the capacity of cattle to learn and remember information about different humans, it may be fair to argue that cattle could possess the ability to learn from heterospecifics (e.g. humans). To the best of our knowledge, it remains unknown if information or behavior can be socially transmitted from humans to cattle (see Nawroth et al., 2019, for farm animals in general). These aspects thus remain to be investigated and highlight the potential to adjust the body language of the human caretakers during handling practices.

6 Cognitive research in an applied setting

General knowledge on the cognitive and sensory capacities of cattle is of utmost importance to develop management routines that are adapted to the needs of the animals and can foster their neural development and thus cognitive capacities. This will help to ensure animals with improved resilience against stressors in later life. Secondly, it is possible to exploit the learning capacities of cattle, for example, to improve safety in handling, to improve both the welfare of the persons handling cattle as well as the welfare of the animals.

6.1 Ontogenetic effects during calf rearing

It is often standard practice to separate the new-born calf and dam immediately, or shortly after birth, in intensive management systems in commercial dairy housing. Calves are raised in individual pens during the milk-feeding period and cows return to the milking herd. In nature and in extensive management systems, calves live in social groups consisting of not only adult cows, but also other calves. Depriving calves of access to similarly aged peers can have a negative impact on their welfare and later productivity. Housing calves individually decreases feed intake and weight gains, but also leads to behavioral problems and cognitive deficits (Costa et al., 2016). Indeed, individual housing of dairy calves can result in measurable learning deficits, indicating impaired behavioral flexibility (Gaillard et al., 2014; Meagher et al., 2015). When confronted with a learning task, calves were given the opportunity to learn a discrimination task, and then re-learn the task when the learned contingencies were reversed. Individually housed and pair-housed calves learned the initial task equally well, but the pair-housed calves were able to adapt more easily when the stimuli were reversed. Pair-housed calves also learned to memorize a novel object after repeated exposures and showed a less anxious response to it, while individually housed calves failed to habituate to it (Gaillard et al., 2014). The social environment was also shown to have an impact on learning in animals (Nicol, 1995). The direct presence of companions may aid associative learning in cattle by decreasing fear. For example, Boissy and Le Neindre (Boissy and Le Neindre, 1990) found that heifers (<2 years old) learned an operant task faster when tested with other heifers than when tested alone. Oppositely, they were slower to learn a task if accompanied by stressed conspecifics, and additionally their latency to approach a bucket containing food increased when exposed to urine from stressed peers (Terlouw et al., 1998). In conclusion, calves raised in isolation exhibit deficient social skills, difficulties in coping with novel situations, and poor learning abilities – all of which may influence the animal's ability to adjust to variable environments on the dairy farm.

6.2 Cow handling and training

Handling animals can be a dangerous job (Pratt et al., 1992) and research indicates that many injuries to both humans and animals in dairy operations happen in relation to manual handling of livestock (Pinzke and Lundqvist, 2007). Areas found to be especially associated with a high injury risk are situations where humans and animals interact directly, such as during moving (Rasmussen et al., 2000) and hoof trimming (Boyle et al., 1997).

Dairy cows are generally docile, but in certain situations (e.g. during moving) they may react unpredictably and thus become hazardous to handle (Grandin, 1999). Fearful cows are not only a threat to human safety, but fear is also detrimental to animal welfare and productivity. Some current approaches focus on the possibility of facilitating positive human–animal interactions to decrease levels of fear during handling. The literature suggests a positive connection between gentle handling and reduction of fear in cattle: positive contact with humans during rearing reduces kicking and facilitates milk let down (Bertenshaw and Rowlinson, 2009), and regular positive contact reduces avoidance distance to humans (Lürzel et al., 2016). In other livestock species,

Figure 5 Dairy cows are generally tranquil and docile, but in some situations where cows interact with humans (e.g. during milking or moving) they may react unpredictably and thus become hazardous to handle. In order to improve such situations, using training regimes from other domestic species might be an option. There is evidence to suggest that training will reduce fear in cattle, while also improving quality of the relationship between handler and cow. Photo: Maria Vilain Rørvang.

such as the horse, there is already considerable work on mapping how the animal reacts to cues from humans and conspecifics in order to avoid risky situations. This includes work on the transmission of safety information among conspecifics in otherwise potentially fearful situations (Rørvang et al., 2015; Rørvang and Christensen, 2018), and using habituation to otherwise fear-eliciting situations (e.g. Christensen et al., 2008) or positive reinforcement training (e.g. McLean et al., 2018). These approaches constitute good potential but have only been used very limited in cattle. One such example is the initiative to train calves to defecate in designated areas, so-called 'toilet training'. Although these studies are the very first steps, cattle show good potential to be toilet trained, which could have implications for future farming practices (Dirksen et al., 2020a,b). There is evidence to suggest that while training will reduce fear in cattle, it may also enhance the quality of the relationship between handler and cow (Ceballos et al., 2018), which may even be generalized to other handlers (Fig. 5).

7 Summary

In this chapter, we summarized the current knowledge on learning and cognition in cattle. Certain aspects, such as cattle's ability to form social categories and the ability to differentiate among individuals, have been thoroughly researched. Other areas, such as cattle's understanding of their physical environment, or social learning abilities, have only been sparsely studied and hence constitute a future research focus. A better understanding of how ontogenetic factors impact cognition and learning is, moreover, a central research focus as it affects the ability of the individual to cope with its environment. The potential to apply novel handling and training regimes based on learning theory also warrants further investigation. The knowledge summarized in this chapter is crucial to better understand how cows perceive and interact with their physical and social environment and can ensure an improved design of husbandry environments and management practices.

8 Where to look for further information

The following articles provide a good overview of the subject:

- Costa, J. H. C., von Keyserlingk, M. A. G. and Weary, D. M. (2016). Invited review: Effects of group housing of dairy calves on behavior, cognition, performance, and health. *Journal of Dairy Science*, 99(4), pp. 2453–2467.
- Marino, L. and Allen, K. (2017). The psychology of cows. *Animal Behavior and Cognition*, 4(4), pp. 474–498.
- Nawroth, C., Langbein, J., Coulon, M., Gabor, V., Oesterwind, S., Benz-Schwarzburg, J., and von Borell, E. (2019). Farm animal cognition–Linking behavior, welfare and ethics. *Frontiers in Veterinary Science*, 6, p. 24.

Key research in this area can be found at the following organizations:

- Animal Welfare program, University of British Columbia (https://awp.landfood.ubc.ca).
- Department of Biosystems and Technology, Swedish University of Agricultural Sciences (https://www.slu.se/en/departments/biosystems-technology/research/animal-environment-and-building-function/).
- Institute of Behavioural Physiology, Leibniz Institute for Farm Animal Biology (https://www.fbn-dummerstorf.de/en/institutes/institute-of-behavioural-physiology/).

Societies and Journals:

- Animal Welfare (https://www.ufaw.org.uk/the-ufaw-journal/animal-welfare).
- Applied Animal Behaviour Science (https://www.journals.elsevier.com/applied-animal-behaviour-science).
- International Society for Applied Ethology (https://www.applied-ethology.org/).
- Journal of Dairy Science (https://www.journalofdairyscience.org/).

- Frontiers in Veterinary Science – Research Topic on Advances and Perspectives in Farm Animal Learning and Cognition (https://www.frontiersin.org/research-topics/6264/advances-and-perspectives-in-farm-animal-learning-and-cognition#overview).

9 References

Bailey, D. W., Rittenhouse, L. R., Hart, R. H. and Richards, R. W. (1989). Characteristics of spatial memory in cattle, *Applied Animal Behaviour Science* 23(4), 331–340. doi: 10.1016/0168-1591(89)90101-9.

Bailey, D. W. and Sims, P. L. (1998). Association of food quality and locations by cattle, *Journal of Range Management* 51(1), 2–6.

Baldwin, B. A. (1977). Ability of goats and calves to distinguish between conspecific urine samples using olfaction, *Applied Animal Ethology* 3(2), 145–150. doi: 10.1016/0304-3762(77)90023-2.

Baldwin, B. A. (1981). Shape discrimination in sheep and calves, *Animal Behaviour* 29(3), 830–834. doi: 10.1016/S0003-3472(81)80017-6.

Barfield, C. H., Tang-Martinez, Z. and Trainer, J. M. (2010). Domestic calves (*Bos taurus*) recognize their own mothers by auditory cues, *Ethology* 97(4), 257–264. doi: 10.1111/j.1439-0310.1994.tb01045.x.

Benham, P. F. J. (1982). Synchronisation of behaviour in grazing cattle, *Applied Animal Ethology* 8(4), 403–404.

Bertenshaw, C. and Rowlinson, P. (2009). Exploring stock managers' perceptions of the human–animal relationship on dairy farms and an association with milk production, *Anthrozoös* 22(1), 59–69. doi: 10.2752/175303708X390473.

Boissy, A. and Le Neindre, P. (1990). Social influences on the reactivity of heifers: implications for learning abilities in operant conditioning, *Applied Animal Behaviour Science* 25(1–2), 149–165. doi: 10.1016/0168-1591(90)90077-Q.

Boivin, X., Garel, J. P., Mante, A. and Le Neindre, P. (1998). Beef calves react differently to different handlers according to the test situation and their previous interactions with their caretaker, *Applied Animal Behaviour Science* 55(3–4), 245–257. doi: 10.1016/S0168-1591(97)00050-6.

Bouissou, M.-F. and Andrieu, S. (1977). Etablissement des relations de dominance-soumission chez les bovins domestiques, *Biology of Behavior* 2, 97–107.

Bouissou, M.-F. and Andrieu, S. (1978). Etablissement des relations préférentielles chez les bovins domestiques, *Behaviour* 64(1–2), 148–157.

Bouissou, M. F., Boissy, A., Le Neindre, P. and Veissier, I. (2001). The social behaviour of cattle. In: Keeling, L. J. and Gonyou, H. W. (Eds) *Social Behaviour in Farm Animals*. Wallingford: CABI Publishing, pp. 113–145. doi: 10.1079/9780851993973.0113.

Boyle, D., Gerberich, S. G., Gibson, R. W., Maldonado, G., Robinson, R. A., Martin, F., Renier, C. and Amandus, H. (1997). Injury from dairy cattle activities, *Epidemiology* 8(1), 37–41. doi: 10.1097/00001648-199701000-00006.

Breuer, K., Hemsworth, P. H. and Coleman, G. J. (2003). The effect of positive or negative handling on the behavioural and physiological responses of nonlactating heifers, *Applied Animal Behaviour Science* 84(1), 3–22. doi: 10.1016/S0168-1591(03)00146-1.

Ceballos, M. C., Sant'Anna, A. C., Boivin, X., Costa, FdO., Carvalhal, M. V. and Paranhos da Costa, M. J. R. (2018). Impact of good practices of handling training on beef cattle welfare and stockpeople attitudes and behaviors, *Livestock Science* 216, 24–31. doi: 10.1016/j.livsci.2018.06.019.

Christensen, J. W., Malmkvist, J., Nielsen, B. L. and Keeling, L. J. (2008). Effects of a calm companion on fear reactions in naive test horses, *Equine Veterinary Journal* 40(1), 46–50. doi: 10.2746/042516408X245171.

Costa, J. H. C., von Keyserlingk, M. A. G. and Weary, D. M. (2016). Invited review: effects of group housing of dairy calves on behavior, cognition, performance, and health, *Journal of Dairy Science* 99(4), 2453-2467. doi: 10.3168/jds.2015-10144.

Coulon, M., Baudoin, C., Abdi, H., Heyman, Y. and Deputte, B. L. (2010). Social behavior and kin discrimination in a mixed group of cloned and non cloned heifers (Bos taurus), *Theriogenology* 74(9), 1596–1603. doi: 10.1016/j.theriogenology.2010.06.031.

Coulon, M., Baudoin, C., Heyman, Y. and Deputte, B. L. (2011). Cattle discriminate between familiar and unfamiliar conspecifics by using only head visual cues, *Animal Cognition* 14(2), 279–290. doi: 10.1007/s10071-010-0361-6.

Coulon, M., Deputte, B. L., Heyman, Y., Delatouce, L., Richard, C. and Baudoin, C. (2007). Visual discrimination by heifers (*Bos taurus*) of their own species, *Journal of Comparative Psychology* 121(2), 198–204. doi: 10.1037/0735-7036.121.2.198.

de Passillé, A. M., Rushen, J., Ladewig, J. and Petherick, C. (1996). 'Dairy calves' discrimination of people based on previous handling, *Journal of Animal Science* 74(5), 969-974. doi: 10.2527/1996.745969x.

Dirksen, N., Langbein, J., Matthews, L., Puppe, B., Elliffe, D. and Schrader, L. (2020a). Conditionability of "voluntary" and "reflexive-like" behaviors, with special reference to elimination behavior in cattle, *Neuroscience and Biobehavioral Reviews* 115, 5-12. doi: 10.1016/j.neubiorev.2020.05.006.

Dirksen, N., Langbein, J., Schrader, L., Puppe, B., Elliffe, D., Siebert, K., Röttgen, V. and Matthews, L. (2020b). How can cattle be toilet trained? Incorporating reflexive behaviours into a behavioural chain, *Animals* 10(10), 1889. doi: 10.3390/ani10101889.

Farm Animal Welfare Council (2013). *The Five Freedoms*. Surbiton, UK.

Finkemeier, M. A., Langbein, J. and Puppe, B. (2018). Personality research in mammalian farm animals: concepts, measures, and relationship to welfare, *Frontiers in Veterinary Science* 5, 131. doi: 10.3389/fvets.2018.00131.

Gabor, V. and Gerken, M. (2014). Shetland ponies (Equus caballus) show quantity discrimination in a matching-to-sample design, *Animal Cognition* 17(6), 1233–1243. doi: 10.1007/s10071-014-0753-0.

Gaillard, C., Meagher, R. K., von Keyserlingk, M. A. and Weary, D. M. (2014). Social housing improves dairy calves' performance in two cognitive tests, *PLoS ONE* 9(2), e90205. doi: 10.1371/journal.pone.0090205.

Galef, B. G. and Laland, K. N. (2005). Social learning in animals: empirical studies and theoretical models, *BioScience*, *BioScience* 55(6), 489. doi: 10.1641/0006-3568(2005)055[0489:SLIAES]2.0.CO;2.

Gilbert, B. J. and Arave, C. W. (1986). Ability of cattle to distinguish among different wavelengths of light, *Journal of Dairy Science* 69(3), 825-832. doi: 10.3168/jds.S0022-0302(86)80472-6.

Grandin, T. (1999). Safe handling of large animals, *Occupational Medicine* 14(2), 195–212.

Griffin, A. S. (2004). Social learning about predators: a review and prospectus, *Learning and Behavior* 32(1), 131–140. doi: 10.3758/BF03196014.

Guzmán, Y. F., Tronson, N. C., Guedea, A., Huh, K. H., Gao, C. and Radulovic, J. (2009). Social modeling of conditioned fear in mice by non-fearful conspecifics, *Behavioural Brain Research* 201(1), 173-178. doi: 10.1016/j.bbr.2009.02.024.

Hagen, K. and Broom, D. M. (2003). Cattle discriminate between individual familiar herd members in a learning experiment, *Applied Animal Behaviour Science* 82(1), 13-28. doi: 10.1016/S0168-1591(03)00053-4.

Heffner, R. S. and Heffner, H. E. (1983). Hearing in large mammals: horses (*Equus caballus*) and cattle (*Bos taurus*), *Behavioral Neuroscience* 97(2), 299-309. doi: 10.1037/0735-7044.97.2.299.

Heffner, R. S. and Heffner, H. E. (1992). Hearing in large mammals: sound-localization acuity in cattle (*Bos taurus*) and goats (*Capra hircus*), *Journal of Comparative Psychology* 106(2), 107-113. doi: 10.1037/0735-7036.106.2.107.

Hemsworth, P. H. (2003). Human-animal interactions in livestock production, *Applied Animal Behaviour Science* 81(3), 185-198. doi: 10.1016/S0168-1591(02)00280-0.

Heyes, C. M. (1994). Social learning in animals: categories and mechanisms, *Biological Reviews of the Cambridge Philosophical Society* 69(2), 207-231. doi: 10.1111/j.1469-185x.1994.tb01506.x.

Hirata, M., Arimoto, C., Hattori, N. and Anzai, H. (2019). Can cattle visually discriminate between green and dead forages at a short distance while moving in the field?, *Animal Cognition* 22(5), 707-718. doi: 10.1007/s10071-019-01268-z.

Hirata, M. and Takeno, N. (2014). Do cattle (*Bos taurus*) retain an association of a visual cue with a food reward for a year?, *Animal Science Journal* 55, 231-244.

Hirata, M., Tomita, C. and Yamada, M. (2016). Use of a maze test to assess spatial learning and memory in cattle: can cattle transverse a complex maze?, *Applied Animal Behaviour Science* 180, 18-25.

Howery, L. D., Bailey, D. W., Ruyle, G. B. and Renken, W. J. (2000). Cattle use visual cues to track food locations, *Applied Animal Behaviour Science* 67(1-2), 1-14. doi: 10.1016/S0168-1591(99)00118-5.

Howery, L. D., Provenza, F. D., Banner, R. E. and Scott, C. B. (1998). Social and environmental factors influence cattle distribution on rangeland, *Applied Animal Behaviour Science* 55(3-4), 231-244. doi: 10.1016/S0168-1591(97)00054-3.

Jensen, M. B. (2012). Behaviour around the time of calving in dairy cows, *Applied Animal Behaviour Science* 139(3-4), 195-202. doi: 10.1016/j.applanim.2012.04.002.

Kovalčik, K. and Kovalčik, M. (1986). Learning ability and memory testing in cattle of different ages, *Applied Animal Behaviour Science* 15(1), 27-29. doi: 10.1016/0168-1591(86)90019-5.

Ksiksi, T. and Laca, E. A. (2002). Cattle do remember locations of preferred food over extended periods, *Asian-Australasian Journal of Animal Sciences* 15(6), 900-904.

Laca, E. A. (1998). Spatial memory and food searching mechanisms of cattle, *Journal of Range Management* 51(4), 370-378.

Lürzel, S., Windschnurer, I., Futschik, A. and Waiblinger, S. (2016). Gentle interactions decrease the fear of humans in dairy heifers independently of early experience of stroking, *Applied Animal Behaviour Science* 178, 16-22. doi: 10.1016/j.applanim.2016.02.012.

Marchant-Forde, J. N., Marchant-Forde, R. M. and Weary, D. M. (2002). Responses of dairy cows and calves to each other's vocalisations after early separation, *Applied Animal Behaviour Science* 78(1), 19-28.

McLean, A. N., McGreevy, P. D. and Christensen, J. W. (2018). *Principles of Learning Theory in Equitation Science*. Available at: https://equitationscience.com/learning-theory.

Meagher, R. K., Daros, R. R., Costa, J. H., von Keyserlingk, M. A., Hötzel, M. J. and Weary, D. M. (2015). Effects of degree and timing of social housing on reversal learning and response to novel objects in dairy calves, *PLoS ONE* 10(8), e0132828. doi: 10.1371/journal.pone.0132828.

Mendl, M., Burman, O. H. P. and Paul, E. S. (2010). An integrative and functional framework for the study of animal emotion and mood, *Proceedings. Biological Sciences* 277(1696), 2895–2904. doi: 10.1098/rspb.2010.0303.

Miller, K. and Wood-Gush, D. G. M. (1991). Some effects of housing on the social behaviour of dairy cows, *Animal Science* 53(3), 271–278. doi: 10.1017/S0003356100020262.

Munksgaard, L., de Passillé, A. M., Rushen, J. and Ladewig, J. (1999). 'Dairy cows' use of colour cues to discriminate between people, *Applied Animal Behaviour Science* 65(1), 1–11. doi: 10.1016/S0168-1591(99)00055-6.

Nawroth, C., Langbein, J., Coulon, M., Gabor, V., Oesterwind, S., Benz-Schwarzburg, J. and von Borell, E. (2019) Farm animal cognition–linking behavior, welfare and ethics, *Frontiers in Veterinary Science* 6, 24. doi: 10.3389/fvets.2019.00024.

Nawroth, C., von Borell, E. and Langbein, J. (2015). Object permanence in the dwarf goat (*Capra aegagrus hircus*): perseveration errors and the tracking of complex movements of hidden objects, *Applied Animal Behaviour Science* 167, 20–26. doi: 10.1016/j.applanim.2015.03.010.

Nicol, C. J. (1995). The social transmission of information and behaviour, *Applied Animal Behaviour Science* 44(2–4), 79–98. doi: 10.1016/0168-1591(95)00607-T.

Olsson, A. and Phelps, E. A. (2007). Social learning of fear, *Nature Neuroscience* 10(9), 1095–1102. doi: 10.1038/nn1968.

Padilla De La Torre, M., Briefer, E. F., Ochocki, B. M., McElligott, A. G. and Reader, T. (2016). Mother-offspring recognition via contact calls in cattle, Bos taurus, *Animal Behaviour* 114, 147–154. doi: 10.1016/j.anbehav.2016.02.004.

Pfister, J. A., Stegelmeier, B. L., Cheney, C. D., Ralphs, M. H. and Gardner, D. R. (2002). Conditioning taste aversions to locoweed (*Oxytropis sericea*) in horses, *Journal of Animal Science* 80(1), 79–83. doi: 10.2527/2002.80179x.

Pinzke, S. and Lundqvist, P. (2007). Occupational accidents in Swedish agriculture, *Agricultural Engineering Research* 13, 159–165.

Pongrácz, P., Miklósi, Á, Kubinyi, E., Gurobi, K., Topál, J. and Csányi, V. (2001). Social learning in dogs: the effect of a human demonstrator on the performance of dogs in a detour task, *Animal Behaviour* 62(6), 1109–1117. doi: 10.1006/anbe.2001.1866.

Pratt, D. S., Marvel, L. H., Darrow, D., Stallones, L., May, J. J. and Jenkins, P. (1992). The dangers of dairy farming: the injury experience of 600 workers followed for two years, *American Journal of Industrial Medicine* 21(5), 637–650. doi: 10.1002/ajim.4700210504.

Ralph, M. H. and Olsen, J. D. (1990). Adverse influence of social facilitation and learning context in training cattle to avoid eating larkspur, *Journal of Animal Science* 68(7), 1944–1952.

Rasmussen, K., Carstensen, O. and Lauritsen, J. M. (2000). Incidence of unintentional injuries in farming based on one year of weekly registration in Danish farms, *American Journal of Industrial Medicine* 38(1), 82–89. doi: 10.1002/1097-0274(200007)38:1<82::AID-AJIM9>3.0.CO;2-Q.

Rehkämper, G. and Görlach, A. (1997). Visual discrimination in adult dairy bulls, *Journal of Dairy Science* 80(8), 1613–1621. doi: 10.3168/jds.S0022-0302(97)76092-2.

Renken, W. J., Howery, L. D., Ruyle, G. B. and Enns, R. M. (2008). Cattle generalise visual cues from the pen to the field to select initial feeding patches, *Applied Animal Behaviour Science* 109(2–4), 128–140. doi: 10.1016/j.applanim.2007.03.014.

Rørvang, M. V. (2018). Maternal behaviour and use of maternity pens in parturient dairy cows PhD thesis. Foulum.

Rørvang, M. V., Ahrendt, L. P. and Christensen, J. W. (2015). A trained demonstrator has a calming effect on naïve horses when crossing a novel surface, *Applied Animal Behaviour Science* 171, 117–120. doi: 10.1016/j.applanim.2015.08.008.

Rørvang, M. V. and Christensen, J. W. (2018). Attenuation of fear through social transmission in groups of same and differently aged horses, *Applied Animal Behaviour Science* 209, 41–46. doi: 10.1016/j.applanim.2018.10.003.

Rørvang, M. V., Christensen, J. W., Ladewig, J. and McLean, A. (2018a). Social learning in horses–fact or fiction?, *Frontiers in Veterinary Science* 5, 212. doi: 10.3389/fvets.2018.00212.

Rørvang, M. V., Herskin, M. S. and Jensen, M. B. (2018c). The motivation-based calving facility: social and cognitive factors influence isolation seeking behaviour of Holstein dairy cows at calving, *PLoS ONE* 13(1), e0191128. doi: 10.1371/journal.pone.0191128.

Rørvang, M. V., Nielsen, B. L., Herskin, M. S. and Jensen, M. B. (2018b). Prepartum maternal behavior of domesticated cattle: a comparison with managed, feral, and wild ungulates, *Frontiers in Veterinary Science* 5, 45. doi: 10.3389/fvets.2018.00045.

Rørvang, M. V., Nielsen, B. L. and McLean, A. N. (2020). Sensory abilities of horses and their importance for equitation science, *Frontiers in Veterinary Science* 7, 633. doi: 10.3389/fvets.2020.00633.

Rushen, J., Taylor, A. A. and de Passillé, A. M. (1999). 'Domestic animals' fear of humans and its effect on their welfare, *Applied Animal Behaviour Science* 65(3), 285–303. doi: 10.1016/S0168-1591(99)00089-1.

Rybarczyk, P., Koba, Y., Rushen, J., Tanida, H. and de Passillé, A. M. (2001). Can cows discriminate people by their faces?, *Applied Animal Behaviour Science* 74(3), 175–189.

Schaeffer, R. G. and Sikes, J. D. (1971). Discrimination learning in dairy calves, *Journal of Dairy Science* 54(6), 893–896. doi: 10.3168/jds.S0022-0302(71)85937-4.

Shettleworth, S. J. (2010). *Cognition, Evolution, and Behavior*. Oxford: Oxford University Press.

Špinka, M. (2006). How important is natural behaviour in animal farming systems?, *Applied Animal Behaviour Science* 100(1–2), 117–128. doi: 10.1016/j.applanim.2006.04.006.

Stoye, S., Porter, M. A. and Stamp Dawkins, M. (2012). Synchronized lying in cattle in relation to time of day, *Livestock Science* 149(1–2), 70–73. doi: 10.1016/j.livsci.2012.06.028.

Taylor, A. A. and Davis, H. (1998). Individual humans as discriminative stimuli for cattle (*Bos taurus*), *Applied Animal Behaviour Science* 58(1–2), 13–21. doi: 10.1016/S0168-1591(97)00061-0.

Terlouw, E. M. C., Boissy, A. and Blinet, P. (1998). Behavioural responses of cattle to the odours of blood and urine from conspecifics and to the odour of faeces from carnivores, *Applied Animal Behaviour Science* 57(1–2), 9–21. doi: 10.1016/S0168-1591(97)00122-6.

Uller, C. and Lewis, J. (2009). Horses (*Equus caballus*) select the greater of two quantities in small numerical contrasts, *Animal Cognition* 12(5), 733–738. doi: 10.1007/s10071-009-0225-0.

Veissier, I. (1993). Observational learning in cattle, *Applied Animal Behaviour Science* 35(3), 235–243. doi: 10.1016/0168-1591(93)90139-G.

von Keyserlingk, M. A. G. and Weary, D. M. (2007). Maternal behavior in cattle, *Hormones and Behavior* 52(1), 106–113. doi: 10.1016/j.yhbeh.2007.03.015.

Whiten, A. (2000). Primate culture and social learning, *Cognitive Science* 24(3), 477–508. doi: 10.1207/s15516709cog2403_6.

Wyatt, T. D. (2003). *Pheromones and Animal Behaviour - Communication by Taste and Smell*. Cambridge: Cambridge University Press.

Chapter 3

Advances in understanding pain and stress in cows

Kenneth M. D. Rutherford and Marie J. Haskell, Scotland's Rural College (SRUC), UK

1 Introduction

Pain and stress are inherent aspects of many situations in the lives of cattle on commercial farms. When measures have not been taken to provide adequate pain relief, pain can be experienced as part of routine farm practices such as castration and disbudding or dehorning in youngstock. Pain is also associated with various diseases, can occur following accidental injuries and may be experienced during parturition. Similarly, there are many aspects of current systems used to manage beef and dairy cattle that may cause stress. Social stress is often a major issue, as cattle are managed in groups which may be larger, more densely stocked or of a different social structure than the conditions cattle have been evolutionarily adapted to (Bouissou et al., 2001). Cattle may be exposed to management procedures that involve periods of isolation, handling and proximity to humans that causes them stress and fear. Under certain circumstances cattle may also experience nutritional stress, when feed availability does not meet the needs of the animal. Feed restriction may also be imposed as part of a management procedure, such as the process of drying-off in dairy cattle, where lactation is ended more or less abruptly, often when the cow is still producing considerable amounts of milk (Fujiwara et al., 2018).

http://dx.doi.org/10.19103/AS.2020.0084.03

Additionally, irrespective of the management system, cattle can be exposed to thermal stress, which may be alleviated or exacerbated by the housing facilities or topography of the grazing land.

Identification of environmental or husbandry causes of pain and stress and the quantification of the associated severity of experience are very important. Interventions can then be designed, or recommendations can be made that will alleviate or reduce this pain and stress.

Pain and stress are affective states. Affective states (or emotions) can be defined as psychological and biological states that have a valence (positive vs. negative) and an arousal (active vs. inactive) element (Mendl and Paul, 2020). As animals evolved to escape from or remove the source of pain or stress, then we can say that affective states cause changes in motivational states. Cognitive processes are also involved in these affective states, as animals are able to remember the stimuli present in pain and stress events and recall them in later similar events (Mendl and Paul, 2020). Assessment of these negative states is central to most modern concepts of animal welfare. The concept that emotion, motivation and cognition all play a role in an animal's experience of, and response to, any event is particularly critical to how we assess and alleviate pain and stress. Critically, however, how the experience 'feels' to the individual is the most important aspect of all.

Therefore, we contend that the most important advances in understanding pain and stress in cattle that have been made–or should be aimed for in the future–relate to our understanding of what particular states mean to the animal itself. Traditionally, assessments of pain and stress have involved quantification of changes in behaviour and in physiological indicators. However, more sophisticated methodologies are required to truly identify the significance of the pain and stress to the individual. The closer we can get to knowing what the individual animal is feeling, the better we can understand the significance of the pain stimuli or stressor to that animal. This is a challenging task and the subject of much scientific debate, but in our view, it is the central issue. It is important, not just because it gives us the fullest account of the animal's welfare, but because in some cases human caretakers may be asked to sacrifice time or money, or production performance in order to remedy putatively pain- or stress-related states. Indeed, in many cases this is a legislative requirement. However, priority should clearly be given to states that are most important from the animal's perspective.

In this chapter we will first provide definitions of pain and stress. In particular, we note the issue that 'stress' as a stand-alone generic term may not be helpful in terms of advancing our understanding of the impact of animal management on animal welfare. For cattle, as for other animals managed by humans, it is necessary to consider the source of 'stress' and to consider the general and specific animal responses to different contexts and events.

Secondly, we will outline the standard methods of assessing pain and stress. We will then present a number of new methods that are being used to assess the experiential aspects, explain their theoretical background and how they allow us to gain a better understanding of the experience of pain and stress in cattle than previous methods with respect to the major sources of pain and stress. Finally, we provide a case study describing a situation where these approaches have led to clear recent improvements in practice that are having real benefits for animals under commercial management.

2 Definitions of pain and stress

2.1 Pain

The question of exactly what pain is has provoked much discussion (Aydede, 2006). Pain is classically defined as: 'An unpleasant sensory and emotional experience associated with, or resembling that associated with, actual or potential tissue damage' (Raja et al., 2020). What is clear from this definition is that pain has different components: a sensory-discriminative part, which relates to the coding of location and intensity of a stimulus threatening a body tissue, and an affective-motivational dimension, which involves the aversive feeling associated with that potential damage and the imperative value given to behavioural pain mitigation strategies (Auvray et al., 2010). Both these dimensions contribute to the unique nature of pain in humans. However, in animals, where the nature of any internal experience is inherently unknown, combinations of behavioural and physiological, and sometimes neurobiological, parameters may be measured in order to make inferences about the pain experience. The term pain can be considered–from a human perspective–to include experiences that range from the insignificant and transitory to the overwhelming. The fundamental challenge in animal pain assessment remains placing particular states correctly along this continuum.

2.2 Stress

As with pain, the term 'stress' is widely used in day-to-day life. But unlike pain, in scientific terms there is no single accepted definition of what stress is. As a consequence, the term is variably applied and can be used to mean the source of a challenge, the experience the animal has as a consequence or the biological changes associated with that experience. Here we take the view that it is helpful to differentiate the source (the *stressor*), the resulting *stress state* (emotional experiences) and the associated *stress response* (Dhabhar and McEwen, 2001). Koolhaas et al. (2011) defined stress as a 'cognitive perception of uncontrollability and/or unpredictability that is expressed in a physiological and behavioural response'. This definition is framed with clear reference to the perceptions of the animal and emphasise that the situations with most meaning (in terms of perceptions of threat) are those where the animal either cannot make anticipatory responses in order to mitigate the challenge, or where responses occur over an extended period because they fail to bring the challenge under control. This means that for acute responses the latency for physiological systems to return to normality seems to be an important indicator of how the challenge is perceived by the animal. The further implication of the definition is that while aversive situations invoke a physiological 'stress' response (e.g. activation of hypothalamic-pituitary-adrenal (HPA) and sympathomedullary (SAM) axes) not all activations of these systems should be considered as stress. Indeed, since the biological function of this activation is to mobilise resources to support behavioural action, many other circumstances also trigger similar responses (the latter are often discussed under the general heading of 'arousal'). Koolhaas et al. (2011) also noted that stressor intensity is also of obvious relevance, and they suggested that mild or transient challenges (i.e. those which are not remotely life threatening) should not be considered as examples of 'stress' even when they are uncontrollable or unpredictable. We also contend that talking generically about 'stress' is not helpful within animal welfare science. While aspects of the stress response may be common across related stressors, to truly understand any given situation from the animal's perspective it is necessary to identify the nature of the stressor and to name the likely associated stress state. For instance, a predator cue could be a stressor that provokes a fear state; deprivation of food acts as a stressor to invoke a hunger state; social defeat acts as a stressor to invoke a social subordination stress state; indeed tissue injury can be considered a stressor that provokes the state of pain. Considered most generally, stressors are real or perceived threats to the animal's immediate survival (e.g. challenges to homeostasis) or long-term survival and reproduction as a consequence of deprivation of resources or immediate physical threat.

The biological response to a stressor can take the form of an autonomic, neuroendocrine, immune or behavioural response, or a combination of these (Moberg, 2000). Behavioural responses are a means of removing or dealing with physical stressors, such as seeking shade to deal with thermal stress or avoiding an aggressive dominant animal (Chebel et al., 2016). The autonomic responses (flight or fight) are short-lived and involve cardiac responses (such as increased heart rate), the adrenal glands and gastrointestinal system. The neuro-endocrine response involves the HPA axis and is a longer-term response. This may affect the metabolic and immune function systems (Elsasser et al., 2000). Over the long-term, unmitigated stress states cause a shift in the allocation of resources from activities such as growth and reproduction to simple maintenance processes. In some cases, animals may exist in a state of 'sub-clinical stress' where the biological cost does not obviously impinge on functioning (Moberg,

2000). In such a state, animals may be under an increased risk for pathology should they be challenged with another stressor (Elsasser et al., 2000; Moberg, 2000). In stress research such a biological cost has been called an allostatic load (McEwen and Wingfield, 2003; McEwen, 2005). Allostatic load is the accumulated cost (the 'wear and tear') that results from repeated physiological alterations (e.g. repeated stress responses in a stress-reactive individual), and/or elevated physiological set points or increased physiological 'effort' required to maintain a set point (McEwan and Stellar, 1993).

3 Foundation methods and the need for new approaches

As noted in the introduction, assessment of pain and stress in an animal welfare context has most often involved the assessment of behavioural and physiological responses to the state under investigation. In pain studies this often involves assessment of how well analgesic provision ameliorates the change, and in stress research since–as discussed above, the concept of stress is best discussed in relation to specific stressors–the 'standard' method of assessing stress varies with the stressor. In both cases, in terms of physiology, assessing level of stress hormones, particularly cortisol, in blood, saliva, hair or faeces (Palme, 2019), is a method that has been used a great deal. Other aspects of biological function–such as assessment of deep body temperature in studies of heat or cold stress (see Hahn, 1999) or changes in heart rate in studies of fear (e.g. Rushen et al., 1999) –have also been used. These studies have allowed clear progress to be made in terms of understanding negative states in cattle and other species, but as discussed by Weary et al. (2017), these can be considered to provide only moderate inference with regard to underlying affective states. For instance, there remains a fundamental issue regarding how the magnitude of the responses traditionally measured should be interpreted in relation to the importance of the experience to the animal itself. The need to be able to quantify or characterise the 'experience' of the animal has become particularly apparent and important in the field of pain research (e.g. Sotocinal et al., 2011) and is becoming more of a goal in assessments of the impact of particular stressors (e.g. Polsky and Von Keyserlingk, 2017).

The study of pain in humans has begun to recognise that the individual's affective-motivational state is central to that individual's experience of pain. This uptake has been largely due to perceived deficiencies in the translational value of existing methods in terms of drug development (i.e. that simple behavioural tests do not reflect the affective component of pain well, which is ultimately what human users want mitigated). Thus, in recent years, the biomedical research field has seen a rapid increase in new methods (or modifications of existing methods) that attempt to assess the 'motivational-affective' component in animal models of pain (Mogil, 2009). Similarly, recent research into stress and stressors relevant to the welfare of managed animals has also begun to use techniques that attempt to quantify the experience of the animal rather than just the changes in endocrinological parameters and behavioural time budgets. These changes have not been driven by the inadequacies revealed by research into pain, but rather reflect an increasing emphasis in animal welfare science on the need to understand the emotional experiences of animals. However, the same underlying principle is applied: that the individual's emotional/motivational experience is a key to its well-being. Many of these methods are not new to animal welfare science and have been used in this field for many years for different purposes. However, the interpretation of the outcomes from the animal's perspective is a new aspect. Additionally, there are other methods that are aimed directly at assessing emotional state and are now being used to assess pain and stress.

4 Sources of pain and stress and foundation measures

Cattle may experience pain and/or stress at various points during their lives. As previously mentioned, it is not useful to talk about stress as a generic phenomenon. While some of the physiological stress responses are similar, there are many sources of stress (stressors) and many corresponding behavioural responses that may be specific to that dealing with that stressor; therefore, the major areas of stress for cattle will be discussed.

4.1 Pain

Management procedures: Management procedures performed on calves has probably received the most attention with regard to pain and its alleviation. Under many cattle production systems calves/youngstock may be subject to procedures such as castration, disbudding/dehorning, tail-docking or methods for identification.

Surgical procedures: Cattle may undergo surgical procedures under farm conditions for various reasons. These include a caesarean section or a surgical resolution of disorders such as displaced abomasum. Although numerically less important, cows are also subject to surgical procedures as part of their use in research, for example, fistulation procedures, where some studies have investigated the benefits of different analgesic regimes (Newby et al., 2014), or shorter duration procedures such as mammary biopsy (Miller-Cushon et al., 2019).

Disease/disorders: Most notably, lameness is seen to occur at a high prevalence in all countries practising modern intensive dairy farming (Rutherford et al., 2009; Solano et al., 2015). While lameness itself is a symptom of a variety of different underlying conditions (including direct physical damage through to infectious sources) many studies have clearly identified pain associated with it. Indeed, lameness itself can be thought of as a guarding behaviour practiced by animals in order to limit their own pain. Studies have found lameness to cause some clear alterations in time budget which appear to reflect underlying motivational trade-offs. Such a change in time budget may manifest as an absolute change or an alteration in behaviour bout frequency and duration. However, across different studies, findings are quite variable, for instance, with regard to parameters like lying time and bout durations.

Parturition: Over the last decade attention has also turned to consider pain experiences relating to giving birth, particularly in cases of dystocia. Initially, this was primarily examining the benefits of pain alleviation to the cow (Laven et al., 2012), but more recently the benefits of analgesia provision for calves following parturition has also been examined (Gladden et al., 2019).

4.2 Thermal stress

Many animals, particularly those that are free-ranging or managed outdoors, are periodically exposed to climatic conditions that they experience as being too hot or too cold. This concept of being 'too hot' or 'too cold' indicates that every animal has a 'thermal comfort zone' (Silanikove, 2000; Kadzere et al., 2002; Kingma et al., 2014), where there are no behavioural or physiological adjustments to maintain its core body temperature in the desired range. However, outside the thermal comfort zone, minimal behavioural or physiological efforts must be applied until the upper and lower limits of the thermoneutral zone (upper and lower critical temperatures) are reached and significant changes in metabolic rate must be made (Yousef, 1985). It is not just ambient temperature that defines the thermoneutral zone (TNZ). Relative humidity, air movement and solar radiation are also contributing factors (Armstrong, 1994).

As mentioned previously, behavioural and physiological mechanisms are used to maintain core body temperature in the desired range. Thus, the standard methods used to assess thermal stress have assessed the characteristic behavioural and physiological responses. In hot conditions, time spent in lying down and feed intake is reduced, while water intake increases (Cook et al., 2007; Schutz et al., 2010; West, 2003). Cows will also make use of resources such as natural or man-made shade or sprinklers (Tucker et al., 2008; Legrand et al., 2011; Schutz et al., 2011; Chen et al., 2013) or move to places where there is more air movement to aid cooling (West, 2003, Haskell et al., 2013). In cold, wet conditions, animals also adjust their posture and time budgets. In cold wet outdoor conditions, cows reduce the time they spend in lying down (Tucker et al., 2007; Webster et al., 2008; Morgan et al., 2009) and seek shelter to reduce exposure to rain or snow (Morgan et al., 2009; van laer et al., 2015). Calves will also 'nest' into bedding as insulation (Lago et al., 2006).

Many studies have used deep body temperature, sweating and respiration rates as measures of heat stress as these parameters increase in hot conditions (Yousef, 1985; Blackshaw and Blackshaw, 1994; Ominski et al., 2002). Changes in body temperature are also used to assess response to cold, as well as changes in hormones indicating physiological responses to increase heat production (e.g. Webster et al., 2008).

4.3 Fear

Fear can be defined as an emotional state induced by the perception of actual danger that threatens the integrity and well-being of the individual. Anxiety is a closely related experience and is a response to potential danger (Boissy, 1995; Forkman et al., 2007). The situations or stimuli that evoke unconditioned or innate fear responses are related to situations associated with threat or danger in that species' evolutionary history. Fear responses are also associated with events or stimuli that had strongly negative or aversive consequences in the animal's own experience (Gray, 1987). Fearful responses are typically expressed as an acute startle response or avoidance behaviour to a stimuli or situation, accompanied by a physiological response that is similar to an acute stress response. This, as well as the fact that fear and fearfulness are important aspects of welfare, means that fear is important as part of this chapter.

There are many situations in our cattle farming and husbandry in which animals may experience stress and fear. However, research into fear in cattle has particularly focussed on the major stimuli and situations that are arguably the most commonly occurring on farms and have economic or welfare consequences for animals. These are human-animal interactions and response to novelty (e.g. see Waiblinger et al., 2006; Forkman et al., 2007 for reviews).

Wild animals typically show a marked avoidance of humans (Price, 1984) and unhandled or naïve cattle show the same response. A number of experimental protocols have been developed that assess the behavioural response to proximity to humans. In beef cattle, the response to the combination of restraint and proximity to humans is typically assessed by scoring the response to restraint in a chute or crush (e.g. Fordyce et al., 1985; Grandin, 1993), the speed that it runs away from this experience (Burrow et al., 1988 Curley et al., 2006a), or their response to handling in yards (Boivin et al., 1994; Le Neindre et al., 1995). In dairy cattle, spontaneous observations of response to humans (e.g. Rushen et al., 1999) and tests are used that assess the animal's tendency to approach or withdraw from a human (e.g. Waiblinger et al., 2003; Waiblinger et al., 2006; Gibbons et al., 2009). Assessments of response to milking (such as the 'step, kick, flinch' test) that observe the behavioural responses to the procedure and associated proximity to humans have been conducted (e.g. Willis, 1983; Breuer et al., 2000). Measures of HPA activation and autonomic nervous system activity have been used to assess the physiological response to handling in beef and dairy cattle (Fell et al., 1999; Hopster et al., 2002; Curley et al., 2006b; King et al., 2006).

There are a number of well-recognised tests for assessing an individual animal's response to a novelty stimulus or a novel situation and the subject has been reviewed extensively for farm animal species (e.g. Forkman et al., 2007; Nielsen, 2020). Some of these tests have been adapted from tests used widely with laboratory animals and adapted for cattle. The 'open field test' or 'novel arena test' involves monitoring the response of the animal to an open, bright, socially isolated space and assessing the behavioural responses. The 'novel object test' involves observing responses to a novel object placed in a test arena (Forkman et al., 2007) and typically involves exposing an individual animal to a novel object (e.g. umbrella: Bourguet et al., 2015; bucket: Lecorps et al., 2018; metal cube: Kilgour et al., 2006). There are a number of variations on this theme including combining the two tests together (van Reenan et al., 2005) or 'startle tests' that assess responses to sudden exposure to an event (e.g. Kilgour et al., 2006; Gibbons et al., 2009).

4.4 Social stress

In the wild situation, cattle live in relatively small family groups in grassland or mixed grassland/woodland habitats. Social groups are stable in composition, with the entry of unfamiliar animals uncommon (Bouissou, 1980). This is in contrast to modern management systems. The typical management system for dairy cattle involves a 'dynamic' social grouping system in which cows are added to the main milking herd as they start lactation and enter a different group at the end of lactation (at 'dry off'). In systems where calving is not clumped or seasonal, and is spread across the year, this creates a situation in which cows can be moved to the larger lactating group on their own–a rather daunting prospect for a young animal naïve to the set-up. Additionally, if the farmer provides different types of feed according to the stage of lactation, pregnancy or growth stage, this will involve managing the cattle in a further set of dynamic sub-groups. Cattle in pasture-based systems tend to be managed around a tighter calving period in the spring to make use of spring grass growth, but they may be managed in very large groups. Beef cattle in intensive management systems such as feedlots may also experience high-stocking densities and large-group sizes. High-stocking densities can increase competition between animals for access to resources such as feed and lying spaces. A lack of space can reduce the opportunities of subordinate animals to show appropriate avoidance or appeasement behaviour towards dominants. Being managed in large groups may hinder the ability of the animals to form stable social groups. However, managing cattle in large groups is very common in both dairy and beef cattle production systems, and much research has been carried out into the effects on the animals.

Many studies have assessed the effects of social stress on responses of the HPA axis. Bouissou (1980) found that close contact between animals during gathering for milking resulted in an increase in weight of adrenal glands. Studies assessing behavioural aspects of overstocking have assessed aggression, and changes in feeding/grazing times and activity. In situations where space is limited, cattle show reduced lying and ruminating times, and increased standing times (Phillips and Rind, 2001; Fregonesi et al., 2007; Krawczel et al., 2012; Fujiwara et al., 2019). Increased aggression is also shown (Phillips and Rind, 2001). Effects are stronger in cows of lower dominance rank compared to those of high rank (e.g. Phillips and Rind, 2001; Huzzey et al., 2006). HPA axis parameters have been assessed, but do not often show effects of stocking density, perhaps as the animal's system adapts to the long-term nature of the social stressor (e.g. Fujiwara et al., 2019).

A number of studies have looked at the response of cattle to overstocking at the feedbunk. These studies have shown that low feeder space per cow results in animals reducing the time they spend at the feed-trough, with shorter feeding times and feeding bouts and eating faster (DeVries et al., 2004; Collings et al., 2011). This suggests that they find the experience of a crowded feedbunk aversive and alter their behaviour to reduce the time spent there.

4.5 Hunger

Cattle are not subjected to severe feed restriction in the way that broiler breeders and sows may be (e.g. D'Eath et al., 2009). As there is typically a direct relationship between feed intake and milk yield for dairy cows and growth for growing beef cattle, it is generally in the farmer's interest to not restrict the amount of feed available. However, there are a few instances when feed quality or quantity is reduced as part of a deliberate management strategy. For dairy cows, one of these is at 'dry-off', a term used to describe the process by which milk production is ceased in the lactating dairy cow in the 6–8 weeks prior to calving. Cows may be dried-off abruptly, which involves stopping milking completely. Some farmers practice a more gradual dry-off by reducing milk yield in the cows by reducing the frequency of milking, or the quantity or quality of feed, or a combination of these two strategies over a few days prior to stopping milking altogether (e.g. Bertulat et al., 2015; Fujiwara et al., 2018).

The practice of 'limit-feeding' in dairy heifers is another example in which the ration is limited in quality. Limit-feeding involves feeding a nutrient-dense diet that is low in roughage. Feed efficiency is higher and feed costs are lower on these diets (e.g. Hoffman et al., 2007). For housed or intensively managed beef and dairy cattle, there may be times when access to feed is limited due to feed-trough design not allowing access to feed for all animals at all times, or when the farmer attempts to reduce wastage by only providing enough feed such that it is all eaten every day (the so-called 'slick bunk' approach: Collings et al., 2011). For cattle on pasture, there may be periods when the availability of pasture does not fulfil their requirements (e.g. Schutz et al., 2006). There is concern that these practices may compromise welfare.

Analysis of feeding behaviour and competition for feed is typically used to assess the response of cattle to feed restriction or restriction in feed access. When subjected to a short period of feed deprivation, cows will spend more time eating and less time lying down once the feed is introduced again (e.g. Metz, 1985), When the period of access to feed is limited by overstocking at the feedbunk, cows will eat faster and spend a shorter time on feeding (Proudfoot et al., 2009; Collings et al., 2011). A number of studies in dairy cattle have shown that aggression is higher at feed-troughs that are crowded and therefore have little space available per cow (e.g. DeVries et al., 2004; Huzzey et al., 2006).

Cortisol analyses are often used to assess the stress responses of cows to the dry-off and, in particular, to the changes in diet quality. Dramatic rises in cortisol are shown in response to abrupt dry-off, particularly in high-yielding cows (>20 kg/day) (Bertulat et al., 2013), but also was shown in a study specifically assessing effects of reduction in feed quality (Odensten et al., 2007).

5 New methods and examples

In addition to the foundation methods, a number of other methods have been developed or adapted to assess pain and stress. Many of these methods are not new, and have been adapted from other fields, or simply have a stronger focus on the aspects that are likely to reflect the emotional experience. These new methods fall into a number of classes. There are new techniques that assess spontaneous behaviour, often with a particular focus on behavioural or facial expression. There are also new behavioural tests, and new applications of existing behavioural tests. Behavioural tests of the outcomes of emotional processing in the brain have also been developed. New physiological assessments have also been developed. In this section, a description of the theoretical background is followed by examples of studies where the method has been used to investigate the experience of animals.

5.1 Behavioural observations

5.1.1 Changes in behavioural priorities

If an animal is willing to engage in a behaviour to gain access to a resource that may cause it injury or pain, it suggests that it highly values that resource. Aggression is one such behaviour. Normally, cattle prefer to avoid engaging in aggression, as it may result in stress and injury. An increased willingness to compete with other animals for a resource shows an increase in the perceived value of the resource. Similarly, it can be argued that forgoing an otherwise valued behaviour also provides an indication of value to the animal.

Heat stress: Studies investigating heat stress have shown that animals are prepared to compete with each other for access to shade (Schutz et al., 2010; Vizzotto et al., 2015) and water (Vizzotto et al., 2015; McDonald et al., 2020) in hot conditions, suggesting that access to these resources is important.

Hunger: As discussed above, a number of studies have shown that aggression increased at the feed-trough when space at the feeder is reduced (Proudfoot et al., 2009; Collings et al., 2011;) and where nutrient density is low, requiring cows to spend a long time in eating (Langford et al., 2011). If cows are prepared to be involved in aggressive interactions, it suggests that gaining access to feed is important to them.

Pain: Disbudding reduced play behaviour a few hours after the procedure but not on the following day (Mintline et al., 2013). Interestingly, local anaesthetic injection to sham animals also reduced play behaviour–backing up other work that suggests this may be aversive to calves in its own right.

Grooming: Lame cows reduce grooming brush use when brushes are far from the feedbunk but not when they are close (Mandel et al., 2018). The authors label grooming behaviour as a 'low-resilience' behaviour, that is, one which is dropped quickly under conditions of challenge (i.e. when resources are low, or as in this case when a competing motivational state takes priority). The aim of the study was to examine whether information on brush use could provide an early detection system for lameness, but the authors concluded this might not be viable as mildly lame cows failed to show this trade-off. However, this highlights the utility of such an approach in terms of understanding meaning (motivational strength) for the animal. Weigele et al. (2018) found that moderately lame cows showed a reduction in frequency of visits to either a concentrate feeder or a grooming brush, and also showed increased lying duration. In relation to the feeder visits, the study found that cows still visited the feeder often enough to achieve their daily feed allowance, whereas sound cows visit more often.

Lying behaviour: Thompson et al. (2019) examined whether lameness affected lying behaviour in pasture grazing cattle. They found no main effect of lameness on lying time but that lameness and rainfall had an interacting effect on this variable. Lying time decreases in cows during periods of rain but decreased less in sound compared to lame cows. The authors speculate that this could be due to lame cows having fewer lying bouts on rainy days due to increased difficulty experienced in transitions between lying and standing.

5.1.2 Context-specific behaviour patterns as indicators of emotional state

Certain behavioural patterns can be considered as expressions of negative emotional states, as they are only observed under conditions of pain and/or distress. They are typically of short duration (e.g. head flicks, tail swishes, bouts of self-grooming) and may be directed at the source of stress (such as behaviours associated with frustration) or at the site on the body which is the source of pain. Vocalisations are another example of these specific behaviours, as they are seen as an 'honest' indicator of emotional state (Watts and Stookey, 2000).

Allogrooming is another behaviour that is linked to the emotional state. There is evidence that allogrooming is a positive experience and it is a behaviour that creates social cohesion in group-living animals. The incidence of allogrooming has been suggested as a welfare indicator. Rises or falls in allogrooming may therefore indicate periods of social stress (Boissy et al., 2007).

Thermal stress: When exposed to wind speed of 1 m/s or more, calves show reactive or 'annoyance' behaviours, such as head-shaking and tail-flicking. Additionally, calves will move away from the source of the wind, suggesting that these behaviours are indicative of a negative experience (Bell et al., 2019).

Social stress: When primiparous animals were mixed with multiparous cows, they performed more social licking of other animals (Phillips and Rind, 2001). Another study compared allogrooming in standard and restricted feeder allowance conditions (Val-Laillet et al., 2009). When feeder space was restricted, less social grooming occurred, particularly in primiparous animals, but partly because the feeder area was where animals had been doing the grooming. This may not tell us a lot about individual experience, but it does suggest that behaviours that may be used to create social bonds, such as allogrooming, may be disrupted by overcrowding.

Hunger: The frequency of vocalisations has been used as an indicator of hunger in a number of studies. In an investigation of how the strategy of reducing feed quality after dry-off affects cows, Valizaheh et al. (2008) compared the responses of cows to two hays varying in digestibility. As expected, they found that all cows vocalised more after they were transferred from the lactation ration to the hay diets, indicating hunger after the transition from the higher-quality diet. However, they also found that cows fed poorer-quality hay vocalised at a higher rate than the group on more digestible hay. In a similar study on the effects of feed quantity, Tucker et al. (2009), investigated the effect of two different planes of nutrition (8 kg dry matter/day vs. 16 kg dry matter/day) given to dairy cows in preparation for dry-off. They found that the cows that were given the lower level of dry matter vocalised at a higher rate than cows on the higher plane of nutrition. Both studies concluded that the cows on these diets were experiencing hunger. Vocalisations are also higher in cows on a limit-feeding regime (Hoffman et al. 2007).

Cows that are food-deprived show behaviours indicative of frustration when they are thwarted from accessing feed. Sandem et al. (2002) found that cows that had been deprived of feed for 7 h showed higher aggression, vocalisations and percentage eye-white when presented with a food bin covered with Perspex compared to cows allowed direct access to the feed. Although these behaviours may be the expression of frustration, the presence of frustration does indicate that hungry cows value access to feed and are disturbed by the thwarting of access.

5.1.3 Qualitative behavioural analysis (QBA)

Qualitative behavioural analysis (QBA) is a method of assessing spontaneous behaviour in a way that does not involve quantifying the classical parameters of duration and frequency of behaviours of interest. QBA assesses the animal's body language or the expressive qualities of the body language of the whole animal (Wemelsfelder, 2007). QBA asks human observers to rate the behaviour they see relative to descriptive terms, such as 'relaxed' or 'agitated' (Wemelsfelder et al., 2000). The use of these terms is thought to allow us to directly capture the animal's experience (Wemelsfelder, 2007). There is some debate over how much the outcome of any qualitative assessment reflects the observer's own interpretation of what the animal is feeling, rather than that of the animal (Wemelsfelder, 2007), but nevertheless, the method has been used in many studies to capture expressive aspects of behaviour that cannot be done with other methods (e.g. Rutherford et al., 2012; Vindevoghel et al., 2019).

Pain: A small trial conducted by de Boyer des Roches et al. (2018) investigated whether an experimental mastitis infection affected cows' emotionality, as assessed using QBA alongside a number of clinical and physiological indicators. QBA identified both a period of negative emotionality associated with the acute phase of the infection and a subsequent period of positive emotionality associated with remission. Behavioural expression of emotionality was also seen to associate with key physiological indicators including body temperature, cortisol levels and circulating markers of inflammation.

5.1.4 Facial expression

The ears and eyes of mammals, as well as the rest of their facial musculature, display a range of expressions. Facial expressions have long been believed to be indicators of the emotional experience of the animal (e.g. Darwin, 1872), but are also believed to have a role in communicating the motivational and emotional state of the individual (Waller and Micheletta, 2013). A number of aspects of facial expression have been examined.

Quantification of facial expression has also been applied to the assessment of pain. The original work in humans identified groups of facial muscles that act together to produce facial expressions. The individual components of muscle movement are known as action units (Ekman and Friesen, 1978). Ear position, whisker change, orbital tightening, and cheek and nose bulge were identified as the facial features that most reliably indicate pain (Langford et al., 2010; Sotocinal et al., 2011).

Although ear posture was used as a facial action unit in the original work in rodents, the use of detailed observations of ear posture has been a major new area of research in cattle in recent years, and is a promising method for the assessment of emotional states. Ruminant species are ideal candidates because of their mobile ears. The initial research involved validation of facial expression and ear posture with behaviour and cardiac responses. Much of this work on ear posture was done in sheep and linked ear posture to emotional state (e.g. Reefman et al., 2009, Désiré et al., 2004; Boissy et al., 2011). Boissy et al. (2011) showed that sheep responded to startling events with asymmetrical ear postures. Other work in that research group and elsewhere has shown that there are transient increases in cardiac activity in these situations, consistent with a startle response (Désiré et al., 2004; Désiré et al., 2006). This work has been extended to cattle.

Pain: Gleerup et al. (2015) examined a number of possible indicators of pain in cows and suggested integrating several of these into a single 'Cow Pain Scale'. Among the indicators found to differentiate cows in pain or not were facial expression (e.g. dilated nostrils and muscle tension around the eyes) and ear position (e.g. backwards and tense). Subsequent work, looking at artificially induced mastitis (de Boyer des Roches et al., 2018) found that while ear position appeared to differentiate control and treated animals, a single indicator of facial expression did not.

5.2 Behavioural tests

While spontaneous behaviour is generally recorded in animals in their normal home environments, there are a number of assessments of pain and stress that involve observing the animal's response in an experimental setting. The types of set-up are quite diverse, but the common underlying principle is that they are all 'asking the animal' about its wants and needs, often by observing behaviour (Dawkins, 2005, 2006), based on the common premise that animal welfare should, in part at least, be evaluated from the animal's point of view (Dawkins, 1990).

The behaviour of an animal at any given time can be considered a consequence of motivational trade-offs dictated by fluctuating levels of various internal and external factors (Toates, 1986). The resulting behaviour can be considered as the current priority in terms of supporting long-term survival and reproduction (in a natural context, where inclusive fitness is a relevant concept). While such considerations are not always strictly relevant for captive-managed animals, the behavioural decisions (not necessarily implying a conscious basis) that animal's make can reveal a lot about their experiences. Various paradigms involving assessments of animal choice, preference and motivational state have been developed (Kirkden and Pajor, 2006).

5.2.1 Preference tests

It is assumed that the choices made by the animal are in their own best interests (Duncan, 1992). Such approaches are not new in relation to many aspects of animal welfare assessment; Hughes and Black first used a choice test in 1973. In the classical preference test, an individual animal is placed in a structure with a neutral starting point and two different resources at two different locations (typically a Y-maze or T-maze). The movement of the animal towards one of the options or resources and consumption or use of that resource is used as an indicator of its preference.

Fear: A study by Pajor et al. (2003) used a preference-testing paradigm to investigate whether cattle can learn to avoid poor handling. The tests indicated that the animals used showed a marked preference for both the control (no handling) situation and for food over being hit and shouted at. However, across the group of cows, no avoidance was shown towards being hit, shouted at or use of a cattle prod, indicating that overall, all treatments were treated as being equal. In trials investigating proposed positive interactions, heifers showed no preference for a gentle stroking treatment versus control.

5.2.2 Motivational and 'trade-off' tests

While preference tests can tell us about how animals rank the various options given to them, they do not provide an indication of the value of that resource. Thus the choice made may represent the 'lesser of evils' where all options reduce welfare or a trivial choice where neither option affects welfare (Duncan, 1992). Motivational tests are used to determine the value that the animal places on access to that resource. In these tests, the animal is asked to pay a 'price' or 'work' for access to the resource (Dawkins, 1990). The level of motivation is inferred from the level of work that the animal is willing to perform to gain access (Matthews and Ladewig, 1994; Schutz et al., 2006). In cattle, increased 'price' or amount of work required is typically imposed by increasing the need for physical activity such as pushing on weighted doors or increasing the distance to be walked (Schutz et al., 2006; Franchi et al., 2019).

Another motivational approach is to recognise that at any one time, an animal could be performing any one of a number of behaviour patterns to fulfil its goals. What behaviour is performed reflects the priority those behaviours have. Pain is just one of many motivational states, which can be considered to compete with other states for final control of muscle function and therefore behaviour. This 'competition' provides a mechanism to assess the significance of any given putative pain state to the animal. This assessment can be via 'free' choices, that is, those made by the animal in its normal home environment, or via 'forced' choices where specific experimental set-ups may be used to reveal underlying motivational strengths.

Motivational trade-off tests are a form of 'forced' choice test, in which a resource with an unknown value (e.g. access to shade) is presented alongside a resource that is known to be valued by the animal (access to food is typically used), but the valued resource is made less attractive than normal, by adding an aversive element to the use of that resource (food is available, but water is sprinkled on the face of the animal when it eats). The choice the animal makes between the novel and 'downgraded' known-value resource gives us an indicator of the relative value of the two resources. Additionally, the intensity of the aversive element may be raised or lowered to discover the point at which the animal switches its choice. The advantage of this testing set-up is that it does not usually require the training for the operant response or 'task' that the typical motivational test requires.

Pain: The use of motivational tests in pain is often aimed at understanding how motivated the animal is to avoid further harm to an injured site. Studies of 'forced' motivational trade-offs have been applied in basic pain science as a way of identifying the strength of pain states (e.g. Fuchs and McNabb, 2012). In animal welfare studies relating to pain such approaches are also starting to be used. Ede et al. (2018) tested calves in a trade-off paradigm where approach to a varying level of milk reward was assessed in calves that had a small volume of saline administered via either intramuscular, intranasal or subcutaneous routes, or that had no treatment. Calves that had the intramuscular injection demonstrated an aversion response by taking a longer latency to approach the milk drinker when only a smaller reward was available. The study shows that the different routes of administration are not the same from the calves' perspective; intranasal and subcutaneous are better than intramuscular. The findings–by testing calves with varying levels of reward–also allow some assessment of the importance of the pain/fear state induced by the intramuscular injection to the calves–that is, they are willing to forego a small reward in order to avoid the potential reoccurrence of the experience, but not forego a large reward.

Hunger: Motivational tests have been used to assess the value that cows place on access to feed. Cows on different food deprivation regimes were asked to walk increasing distances for a food reward in a study by Schutz et al. (2006). A nice refinement in this study was that the amount of energy in the food reward was calculated to match the amount of energy used in walking to get to the reward. The results showed that there was a direct relationship between the food deprivation level and the distance that the cows were willing to walk. However, this was only true for lactating cows, as there was no such relationship for late pregnancy but non-lactating cows.

A weighted push-gate was used by Franchi et al. (2019) in a study assessing feeding motivation during drying-off. The motivation to push the weighted panel to get access to concentrate was compared between a group of cows that were receiving a reduced-energy dry-off diet and a control group receiving the standard lactation diet. Cows on the reduced-energy diet had higher maximum weight limits for pushing the gate and had lower latencies to use the push-door than cows on the standard diet.

Similarly, a weighted push-door was used in a study to assess the motivation of limit-fed heifers to gain access to roughage (straw). Heifers on a high nutrient density were more motivated to gain access to the straw than heifers on a diet with a higher fibre content (Greter et al., 2015).

Although the author did not interpret the data in this way, the study by Metz (1985) showing that cows will decrease lying in favour of eating when food-deprived can be reinterpreted as cows trading-off lying for standing. There may be many other studies which may be reinterpreted in this way.

Thermal stress: A study by Schutz et al. (2008) used a trade-off test to investigate the motivation of dairy cattle to use shade in hot weather. It is well-recognised that cows value lying time (e.g. Jensen et al., 2005), and so the value of shade was assessed by giving cows the choice of lying down or using the shade at various levels of lying deprivation. The study showed that in hot conditions, cows that had been deprived of lying for the maximum level (12 h) chose to use the shade, indicating the cows highly value access to shade in these conditions.

Social stress: Studies discussed above show that subordinate animals will alter their feeding behaviours when the space at the feeder is crowded, and suggest that these animals find the experience of a crowded feedbunk aversive. This is reinforced by studies that looked at the willingness of dairy cows to feed next to a dominant individual. In a trade-off study, Rioja-Lang et al. (2009) showed that subordinate dairy cows would rather eat a low-quality feed than eat next to a dominant animal. This was at a feeder-space allowance of 60 cm/cow. A second study showed that the feeder-space allowance must be 60 cm/cow before subordinate cows are willing to voluntarily stand next to a dominant animal at the feed-trough (Rioja-Lang et al., 2012).

5.2.3 Conditioned place preference/aversion tests

A variant of motivational testing is the use of tests of conditioned place preference (CCP) or aversion. These tests are based on the principle that positive or negative aversive states play an important role, not just in dictating immediate reactions to events, but also in providing learning cues that affect future behavioural reactions. In the simplest terms possible, this means that animals will show preferences for cues or environments where previous positive events took place and aversion responses to those associated with negative states. In fundamental pain research these approaches have shown substantial utility in terms of assessing the meaning of pain states associated with particular interventions and also the benefits–to the animal–of analgesic provision (Mogil, 2009). As with the use of motivational-based tests, pain assessment in an animal welfare context has also recently seen examples of this form of testing. Applications to other forms of stressor have not been seen but the methodology would equally well apply.

Pain: Ede et al. (2019b) examined the affective component of pain associated with hot-iron disbudding using a conditioned place preference/aversion paradigm. Calves were disbudded with local anaesthetic only on one horn bud or with local plus analgesic on the other horn bud. The two sides were done on different occasions and side of procedure was paired with a visually distinct different location in a test apparatus. The benefits of reducing the affective dimension of pain from providing the NSAID meloxicam was demonstrated by the fact that calves spent more time in the location associated with local plus NSAID as opposed to local alone, that is, they demonstrated a preference for the location associated with analgesia/an aversion for the place associated with local anaesthetic treatment only. The study also demonstrated differences between two NSAID analgesics: provision of the shorter-acting ketoprofen actually produced an aversion response.

Adcock and Tucker (2020) used a CCP paradigm (associating treatments with location and visual pattern cues) to examine the response of disbudded calves to analgesia (a cornual nerve block) either on the day of the procedure or 20 days afterwards. Place preference responses were compared to sham handled calves. There was only weak evidence for a CCP effect on the day of the procedure (a statistical tendency for disbudded calves to spend more than 50% of their time in the area associated with analgesia, but no overall difference between disbudded and control calves). However, 20 days post-procedure there was a difference between disbudded and control calves: the latter group showed an avoidance response to the area associated with analgesia whereas the disbudded group did not. This appears to demonstrate that pain-free calves perceive some aversive properties of the cornual nerve block but that disbudded animals trade-off this negative experience with some putative additional positive properties of the treatment, that is, relief from ongoing pain.

5.2.4 Cognitive bias tests

There is evidence from both humans and other animals that the emotional state of the individual will influence its response to unfamiliar stimuli and to potentially threatening stimuli, and in the memory storage and recall of these stimuli. The changes in responses due to emotional state are known as cognitive biases (see Paul et al., 2005 for a review). Methodologies to assess two of these processes have been developed for use in farm animals. In judgement bias tests, animals are taught that certain outcomes are associated with particular stimuli and animals learn to make appropriate approach or avoidance responses. Then an intermediate or ambiguous stimulus is presented. If an animal is in a positive emotional state, it will respond as if the intermediate predicts a positive outcome. However, animals in negative emotional states (which are often induced as part of the experiment) will show a less positive response. In attention bias trials, the time animals spend paying attention to potentially threatening stimuli is assessed. Animals in anxious or negative states will spend more time attending to the fear-inducing stimuli (Lee et al., 2018; Monk et al., 2018). Both cognitive bias and attention bias systems have been used to assess emotional state in cattle.

Pain: Neave et al. (2013) used a visual go/no-go discrimination task (colours presented on a video screen) to assess calf judgment bias before or after disbudding (with prior sedation and local anaesthesia). Following disbudding, calves showed a reduction in approach responses to ambiguous screen colours compared to their responses before disbudding. This negative bias was seen at both 6 h and 22 h after disbudding. Responses to unambiguous positive or negative cues were unchanged. In a follow-up study, Lecorps et al. (2019) used a novel judgment bias task to again assess calf responses following disbudding. The study found again that at 6 h after the procedure calves showed evidence of a negative judgment bias. However, in contrast to Neave et al. (2013) they (i) found evidence of reduced response to an unambiguous positive cue (which they suggest might indicate anhedonia) and (ii) found that the negative judgment bias was no longer evident 22 h post procedure.

Fear of humans: Calves which showed the strongest fear responses in a novel object and a human approach test also showed the most pessimistic responses in a judgement bias test (Lecorps et al., 2018), suggesting that understanding individual differences between animals is important. Although traditional tests such as flight speed and crush score are used to assess fear of humans, little is known about how these 'traits' relate to transient emotional state such as fear and anxiety. Lee et al. (2018) investigated the relationship between crush score, flight speed and anxiety. Cattle were scored for crush score and flight speed. One group of treatment animals were then given a pharmacological intervention to induce anxiety or reduce anxiety. The results suggested that anxiogenic drugs did not affect crush score or flight speed. This suggests that anxiety is not an emotional element that drives an animal's response in the crush or in the fleeing from it (Lee et al., 2018). As anxiety is defined as a response to potential threat this may be true, given that the animals directly experienced the restraint and human handling. It may be that the response is an expression of fear, or that crush score and flight speed measure general agitation (Lee et al., 2018).

5.2.5 Laterality

In the context of animal form and function, lateralisation refers to the fact that the left and the right hemispheres of the brain are specialised to process information from environmental stimuli in different ways. The left side of the brain is responsible for processing positive or predictable activities, while the right side is responsible for processing of new or challenging stimuli (Rogers, 2010). From an evolutionary point of view, this means that the right hemisphere is involved in rapid responses and detection of predators, and the left with considered responses and processing of prey or food (Rogers, 2002). Because of a crossover in the optic chiasm, visual signals from the left eye (or left visual field) enter the right hemisphere of the brain, while signals to the right eye enter the left hemisphere (Rogers, 2010). This means that recording whether an individual chooses to look at an object or individual with the left or right eye can tell us about whether the animal views that stimulus as being negative or positive. A number of studies have observed how cattle view stimuli to determine whether they are characterising them as being positive or negative. The aim of the tests is to assess whether the animal chooses to view the novel stimuli through its left or right eye, thus giving some clue to whether that animal perceives the stimuli as being challenging or familiar. The tests typically involve presenting the animal with a single object directly in front of it and/or using two objects equidistant to the animal's left and right and assessing which eye is used preferentially.

Pain: In relation to pain, Kappel et al. (2017) examined whether lameness affected the laterality of behavioural responses to various novel objects (discussed below). They found that while behaviour in relation to novel objects did show laterality this was not related to the degree of lameness shown by the individual cow.

Fear of humans: A number of studies have investigated the phenomenon of lateralisation to assess the response of dairy cattle to humans. A study by Phillips et al. (2015) used a set-up in which cows walking down a track had to pass an experimenter in clothes atypical to those normally worn by staff and students on that farm or an experimenter wearing the familiar blue overalls. Most cattle passed by on the right side of the experimenter in unfamiliar clothes thus using the left eye/right hemisphere, while they passed the experimenter in familiar clothes on the left, using the right eye/left hemisphere. This suggests that the unfamiliar person is being perceived as being novel and potentially threatening, while the person in the familiar clothes is being treated as more 'normal' (Phillips et al., 2015). In a similar study assessing cattle in groups, the animals were more likely to move past the experimenter in a path that meant their left eye was oriented towards them (Robins and Phillips, 2010). A further study provided additional validation, showing that cows that used the left eye to view a novel human also showed behaviours indicative of anxiety (Goma et al., 2018).

Fear of novelty: The phenomenon of lateralisation has also been used to investigate the experiential aspects of the response to novelty. In a study by Kappel et al. (2017), two identical novel objects were hung on either side of a passageway and the cows were allowed to view and approach the objects. More cows looked at the novel objects by turning their heads towards the right object, thus viewing it with their right eyes. As the theory suggests that negative stimuli are more likely to be viewed through the left eye, this right bias suggests that the objects were being viewed as positive. They approached without hesitation, also indicating a lack of fear.

5.4 Physiology

5.4.1 Thermography

Recent years have seen new developments in the application of thermography in animal studies, largely due to improving technology and the availability of equipment at lower cost. Thermography is primarily used to aid clinical diagnosis (i.e. identify sources of a problem) and to assess the extent and severity of inflammatory responses associated with an injury or pathology. As with other physiological measures it may be expected that this provides a 'ball-park' indication of pain experiences but the exact relationship between any given degree of inflammatory reaction and central pain experiences is uncertain. However, the use of thermal imaging has also been proposed as a way of more directly assessing an animal's reaction to a stressor. When the sympathetic nervous system is triggered, blood moves from the periphery to the major organs, in the classic 'flight or fight' response. In areas of the skin at the periphery, which have good blood supply, such as the ears, a short drop in temperature can be observed in an acutely frightening situation, due to blood being directed to the internal organs (Stewart et al., 2005b). However, following this transient reaction, the HPA axis activation causes increased heat production, and an increase in temperature is expected (Schaefer et al., 2002).

Thermal stress: In an experiment investigating responses of cattle to cold, wet conditions, thermographic images taken from the eyes of cattle exposed showed that these peripheral temperatures were lower in cows exposed to wind and rain compared to those housed indoors (Webster et al., 2008).

Fear: As part of a trial validating thermography as a measure of stress, Stewart et al. (2005a) gave cows an ACTH challenge and social isolation as a challenge and assessed eye temperature. As part of the trial, they inserted catheters into animals on two separate occasions. Interestingly, eye temperature was high after the second catheterisation event, suggesting that the animals were anticipating experiencing the unpleasant effect for a second time, suggesting that cognitive anticipation of the experience triggers the response.

5.4.2 Heart rate variability

Heart rate variability (HRV) is defined as the measured variation of the length of the inter-beat intervals and represents the balance between the heart-brain interactions and the dynamics of the sympathetic and para-sympathetic nervous systems. Modulations in each of these systems results in a dynamic physiological control system that is ready to respond to challenge, and is therefore never static in a healthy organism, resulting in constant variation in inter-beat interval (Shaffer et al., 2014). Although some studies have used changes in heart rate as a measure of stress, changes in heart rate are difficult to interpret, because they may result from changes in input from both the sympathetic and parasympathetic inputs. Heart rate variability, however, is seen as a good indicator of psychological stress in humans (Von Borell et al., 2007) and is increasingly being used in animals to assess response to challenge and to positive experiences.

Pain: Pain has been shown to affect HRV parameters in a number of studies. In a very small trial, Frondelius et al. (2015) identified a difference in one measure of heart rate variability between cows that were or were not provided with a NSAID following a minor surgical procedure. Kovács et al. (2015a) compared various parameters relating to HRV in dairy cows that were assessed by locomotion scoring to be either lame or sound. They assessed HRV measures during periods when cows were lying and found clear differences between lame and sound animals (whereas no differences were seen in faecal glucocorticoid metabolite levels). Byrd et al. (2019) found that one HRV parameter differentiated between calves that were disbudded with pain relief and control animals but did not differentiate either of these treatments from a group of calves disbudded without pain relief. Jimenez et al. (2019) found no difference in various HRV parameters between calves that had a local anaesthetic nerve block injection with or without a prior topical anaesthetic. This further substantiated other findings suggesting that such injections are actively aversive to the calves and that the topical treatment did not mitigate the pain.

Fear of humans: Kovács et al. (2015b) investigated heart rate and heart rate variability in dairy cattle in two tests. The first was a temperament test similar to that used in beef cattle, that gave a score to each animal based on the response to restraint (for fitting of the heart rate monitoring equipment) and the second one was a score of the behavioural response to human approach. The study showed that animals with more excitable responses in the temperament test had lower (more unfavourable) HRV values while being milked, but also when they were lying undisturbed. Cows with higher responses to an approaching human had lower HRV and HR values. These results suggest that animals showing higher levels of fear at milking also appear to have higher baseline levels of stress.

5.4.3 Eye-white

The eyes are a part of the face that are important to the expression of emotion. The use of the percentage of white seen in the eye has become more widely used as interest in the quantification of emotional states has increased. Sandem et al. (2002) first suggested that the amount of white seen in the eye might be used as an indicator of emotional state in cattle. So although it is a behavioural response, the opening of the eyes wide is likely under the control of the sympathetic nervous system, as greater eye opening will enhance visual detection during the flight or fight response. This opening of the eye exposes more of the white sclera of the eye around the pupil and iris, and the amount of white sclera as a proportion of total visible eye is assessed using this method.

Fear: This approach was used by Battini et al. (2019) in a study with dairy cattle. They found that cattle showed increased levels of eye-white and more 'forward' ear posture when being approached by an unfamiliar person compared to a neutral situation.

Hunger: Cows that are food-deprived show behaviours indicative of frustration when they are thwarted from accessing feed. Sandem et al. (2002) found that cows that had been deprived of feed for 7 h showed higher aggression, vocalisations and eye-white when presented with a food bin covered with Perspex than cows allowed direct access to the feed. Although these behaviours may be the expression of frustration, the presence of frustration does indicate that hungry cows value access to feed and are disturbed by the thwarting of access.

5.5 Pain-specific indicators

So far, the indicators discussed have been those that have application across the assessment of a number of different sources of stress, including pain as a stressor. However, some measures can be considered pain-specific in that they specifically aim to target the measurement of some functions of the nociceptive or pain system. The two most prominent of these are the assessment of physiological pain biomarkers and the measurement of sensory response thresholds (algometry).

Physiological biomarkers: Although almost any biological trait could be considered as a pain 'biomarker', in reality most studies that use this term restrict it to physiological variables. Often it may include classic stress response markers such as cortisol, but the ideal pain biomarker would be one that was specific to pain. The substance most commonly considered in this area is substance P. Substance P is a neurotransmitter that plays a key role in the transmission of nociceptive information from the periphery to the spinal cord. A number of cattle studies have demonstrated alteration in substance P levels associated with veterinary procedures, operations and disease (Allen et al., 2013; Kleinhenz et al., 2017; Tschoner et al., 2018).

Algometry: Assessment of nociceptive thresholds has also been a common feature of pain studies. Lameness is associated with evidence of lowered thermal response threshold in dairy cows, with evidence of modulation of thresholds in the opposite direction by provision of pain relief (Whay et al., 1998, 2005). Artificially induced mastitis has been associated with both signs of hyperalgesia (Fitzpatrick et al., 2013) and hypoalgesia (Rasmussen et al., 2011). Calf disbudding causes long-lasting hyperalgesia around the site of the injury (Adcock and Tucker, 2018).

In both cases (biomarkers and algometry), it is important to note that alterations in either levels of substance P or mechanical or thermal thresholds are not in themselves indicative of pain experiences (as opposed to informing on nociceptive mechanisms that feed information to the brain for integration that ultimately produces the felt experience of pain), despite the fact that some studies do incorrectly make claims about pain on the basis of such assessments of nociceptive function. That said, alterations to peripheral sensitivity may have relevance for animal welfare in cases where reduced thresholds lead animals to experience increased discomfort or pain as a consequence of environmental stimulation. A further issue is that, as yet, we have a limited understanding of what meaning such changes have for animals in practical contexts. It remains possible that some level of change in sensitivity may be more or less meaningless to the individual animals or alternatively (e.g. where allodynia is present) could represent a severe imposition on their long-term emotional state. This issue is particularly relevant in relation to signs of altered sensitivity that persist following tissue healing. For instance, Casoni et al. (2019) recently found evidence of chronic sensitisation in calves following disbudding. However, such studies should perhaps be taken as providing grounds for studies that investigate the consequences of emotional experience of this for the animal, rather than–in themselves–evidence for a negative experience.

5.6 Conclusion

This discussion of many individual studies serves to illustrate the diversity of new tests and new applications of older methods that have been used to investigate issues around pain and stress in cattle. The common link in this diversity is the aim of better understanding the experience of the animal based on the premise that if we understand the feelings and experience, we can better understand its welfare.

One of the major advances that these studies offer is a more accurate view on 'what matters' and 'how much it matters'. Many of the findings of these experiments provide back-up for the findings of the foundation studies, but by looking at them in a different way. Cattle find over-crowded housing and feeding areas aversive, and they find abrupt dry-off and disbudding painful (Proudfoot et al., 2009; Neave et al., 2013; Franchi et al., 2019). Some of the studies show unexpected results–cattle find all negative handling experiences equally aversive (Pajor et al., 2003), anxiety is not a feature of response to handling (Lee et al., 2018) and cows may respond to novelty as a positive experience (Kappel et al., 2017). These studies show that these testing methods provide new information that can reinforce or challenge current understanding of the experience and needs of cattle.

Measuring indicators of emotional state allow us an understanding of the 'ranking' of resources, such as diet types for hungry cows (e.g. Valizaheh et al., 2008; Tucker et al., 2009). Assessments of emotional response can also provide information on the 'thresholds' at which stressors become important such as proximity to dominant animals at feeding or levels of wind or heat (Schutz et al., 2008; Rioja-Lang et al., 2009, 2012, Bell et al., 2019). They can also pinpoint specific time-points within a painful or stressful event, when the experience is the most aversive, such as during an episode of mastitis (de Boyes des Roches et al., 2018) or after disbudding (Neave et al., 2013).

The major issue that the use of these techniques illustrates is the change in mindset in our attitudes towards animals and what we can ask about them with careful experimentation. The study of emotional state and experience has been legitimised and these techniques are examples of the methods that allow us to ask questions about individual experience. As these experiences are central to the animal's welfare, these advances can provide the evidence to bring about change in management and husbandry to benefit the welfare of cattle.

6 Future research

It is likely that we will see these methods applied to more of the situations in which cattle are likely to experience stress and pain. These include: (1) the role of social observation as a source of transmission. Rat studies show that pain experiences can be contagious, as observation of a conspecific in pain alters pain processing in the observer (e.g. Li et al. (2014); (2) the nature of the physical environment. Laboratory animal studies show that pain experiences may be altered by provision of better quality environments (e.g. Gabriel et al. (2010); (3) early life experience. Events in the early life, including prenatal life, of animals can modify pain and stress experiences throughout life; (4) relationships between affective states and other aspects of biology such as the microbiome, neuroglia and so on. Finally, while this chapter focussed on classical considerations of negative welfare states, that work is increasingly viewed within the wider focus on positive welfare. It remains to be seen how indicators of positive and negative welfare relate or trade-off in animals.

It is likely that future trends will see more application of automated measurement techniques (O'Leary et al., 2020) and application of computing algorithms (e.g. machine learning) to interpret the resulting data. Such approaches are already being applied and will likely develop rapidly in the years to come. However, in this area, as in more traditional forms of assessment, the fundamental issue remains understanding the relevance of different states to the animals concerned. This goal has two benefits. First, a deeper understanding of animal emotional states adds weight to concerns about welfare. Secondly, it allows for the efforts to improve practice (whether through more research, or human behaviour change) to be focussed on the areas where need is greatest.

7 Case study: disbudding in calves

Within the topic areas covered in this chapter, perhaps the best example of research contributing to practical improvements in animal welfare on farm relates to calf procedures. Surveys of cattle vets and farmers (Huxley and Whay, 2006; Remnant et al., 2017) show that disbudding/dehorning are rated as highly painful (e.g. 7/10 on a scale from 'no pain' to the 'worst pain imaginable'). This perception is in line with (and may to some extent have been dictated by) a large body of work which has clearly demonstrated pain and stress associated with the procedure, and how these can be reduced by appropriate pain relief treatments (Herskin and Nielsen, 2018; Winder et al., 2018). These studies have included a wide variety of assessment approaches, from the quantification of behaviour (Faulkner and Weary, 2000), stress physiology (Sutherland et al., 2002; Stewart et al., 2008), production parameters (Bates et al., 2015), nociceptive response thresholds (Adcock and Tucker, 2018), pain biomarkers (Kleinhenz et al., 2017), changes in microbiota (Mir et al., 2019), assessment of brain activity (Gibson et al., 2007) to–more recently, as outlined above–assessment of emotional/motivational affects (Ede et al., 2018, 2019a; Lecorps et al., 2019; Adcock and Tucker, 2020). The results of these arrays of studies are in general agreement about the methods of provision of pain alleviation to calves. Furthermore, some of the more recent studies (Adcock and Tucker, 2020; Lecorps et al., 2019) have started to provide a picture of the possible duration of different types of impact on the animal; for instance, that judgment bias alterations seem relatively short-lived (Lecorps et al., 2019) but that place preference response suggest ongoing pain after 20 days (Adcock and Tucker, 2020).

This body of work as a whole has contributed to the current situation where practices on farm are starting to change. In the UK, the British Veterinary Association and the British Cattle Veterinary Association (BVA/BCVA, 2017) issued a clear policy statement supporting the use of long-acting analgesics in addition to a local anaesthetic. Additionally, the UK Farm Animal Welfare Council (FAWC) (2019) also issued strong guidance along the same lines. Survey studies from the UK, Canada and Finland (Hokkanen et al., 2015; Remnant et al., 2017; Moggy et al., 2017; Winder et al., 2016) clearly show that practices are changing with increasing use of local anaesthesia and ongoing systemic analgesia. In addition, many producer groups (e.g. retailers or milk buyers) have started to add this as a requirement. This is clear evidence of the impact that the years of research into fundamental mechanisms coupled with assessments of emotional experience have contributed to the changing practices on farm.

8 Summary

There are many aspects of cattle husbandry systems that have the potential to cause pain and stress. In some cases, procedures such as disbudding or dehorning in calves and abrupt drying-off are part of the standard management practice on farms. Other forms of stress, such as thermal stress and pain due to accidental injuries, are less under the control of the farmer. Whatever the nature of the source of pain and stress, there can be a major negative impact on the animal's welfare.

Understanding the mechanisms and measurement outcomes has been a major undertaking in animal welfare science in the past. These foundation studies relating to pain and stress in cattle have contributed a great deal to our understanding of how to assess and resolve important animal welfare issues. Many of these studies have assessed behavioural changes as indicators of pain and stress, as well as changes in the HPA axis and autonomic nervous system outputs to assess stress and changes in neurophysiological outcomes to assess pain and major changes in behaviour to assess both.

However, if animal welfare is all about animal feelings (Dawkins, 1990; Duncan, 1992) then we should be measuring parameters that allow stronger inferences about these states to be drawn. The lack of progress in drug development in pain research has given further emphasis to this endeavour. Thus the major new advances in the fields of pain and stress research in all animals have been in the development and use of methods to assess the subjective experience of the individual animals. Some of the methods have been used for many years, but the new usage involves a particular focus on animal experience.

Many of these have been applied in cattle to assess the individual experience of pain and stress. The new assessments include assessments of spontaneous behaviour (observations of context-specific 'indicator' behaviours, changes in motivational priorities and facial expression), behavioural tests of preference and motivation that focus on stimuli that may affect experience, tests of emotionality and new methods that assess physiological responses that are more directly linked to psychological experience than physiological functioning.

What these experiments have shown is first that animal experience can be quantified using careful and well-designed experiments. Secondly, the use of these techniques gives a new perspective on the animal's experience, which may or may not match the results of the traditional measures. Some of the results are surprising, which indicates that this perspective was important to explore. The use of these new approaches has also allowed us to more accurately pinpoint the threshold values at which the experience of heat, cold and over-crowding starts to be perceived as aversive. The value cattle place on resources can also be more easily ranked to allow priorities to be quantified. It can also show the important time-points within a procedure that are the most painful. It is concluded that future methods will go further in assessing the animal's experience, which is ultimately the best indicator of its welfare. The improvements will be driven by advancement in technology, data analysis and the use of new methods. The results can be used to identify the aspects of the environment that cause the most stress and pain, which will allow farmers and others involved in the rearing and care of cattle the information to enable welfare improvements.

9 Where to look for further information

The book entitled 'Biological response to stress: implications for animal welfare' edited by G. P. Moberg and J. A. Mench (2000, CAB International, Wallingford) is a highly cited and fundamental text.

A number of academic journals also publish research on pain and stress in cattle. These include:

- Applied Animal Behaviour Science: https://www.journals.elsevier.com/applied-animal-behaviour-science/.
- Journal of Dairy Science: https://www.journalofdairyscience.org/.
- PLoS ONE: https://journals.plos.org/plosone/.

Websites from research groups and drug companies working in these areas also contain some relevant material. The following are some examples:

- https://www.fawec.org/en/.
- https://www.farmanimalwellbeing.com/.
- https://www.animalwelfarescience.com/.

10 References

Adcock, S. J. J. and Tucker, C. B. (2018). The effect of disbudding age on healing and pain sensitivity in dairy calves. *J. Dairy Sci.* 101(11), 10361–10373.

Adcock, S. J. J. and Tucker, C. B. (2020). Conditioned place preference reveals ongoing pain in calves 3 weeks after disbudding. *Sci. Rep.* 10(1), 3849.

Allen, K. A., Coetze, J. F., Edwards-Callawy, L. N., Glynn, H., Dockweiler, J., KuKanich, B., Lin, H., Wang, C., Fraccaro, E., Jones, M. and Bergamasco, L. (2013). The effect of timing of oral meloxicam administration on physiological responses in calves after cautery dehorning with local anesthesia. *J. Dairy Sci.* 96, 5194–5205.

Armstrong, D. V. (1994). Heat stress interaction with shade and cooling. *J. Dairy Sci.* 77(7), 2044–2050.

Auvray, M., Myin, E. and Spence, C. (2010). The sensory-discriminative and affective-motivational aspects of pain. *Neurosci. Biobehav. Rev.* 34(2), 214–223.

Aydede, M. (Ed.) (2006). *Pain: New Essays on Its Nature and the Methodology of Its Study*. MIT Press, Cambridge, MA.

Bates, A. J., Eder, P. and Laven, R. A. (2015). Effect of analgesia and anti-inflammatory treatment on weight gain and milk intake of dairy calves after disbudding. *N. Z. Vet. J.* 63(3), 153–157.

Battini, M., Agostini, A. and Mattiello, S. (2019). Understanding cows' emotions on farm: are eye white and ear posture reliable indicators? *Animals* 9(8), 477.

Bell, D. J., Haskell, M., Macrae, A., Mitchell, M., Jennings, A. and Mason, C. (2019). Behavioural responses of dairy calves to temperature and wind. *Proceedings of the 28th Annual Congress of the ISAE*, Bergen, Norway, 5-9 August, 2019.

Bertulat, S., Fischer-Tenhagen, C. and Heuwieser, W. (2015). A survey of drying-off practices on commercial dairy farms in northern Germany and a comparison to science-based recommendations. *Vet. Rec. Open* 2(1), e000068.

Bertulat, S., Fischer-Tenhagen, C., Suthar, V., Möstl, E., Isaka, N. and Heuwieser, W. (2013). Measurement of fecal glucocorticoid metabolites and evaluation of udder characteristics to estimate stress after sudden dry-off in dairy cows with different milk yields. *J. Dairy Sci.* 96(6), 3774–3787.

Blackshaw, J. K. and Blackshaw, A. W. (1994). Heat stress in cattle and the effect of shade on production and behaviour: a review. *Aust. J. Exp. Agric.* 34(2), 285–295.

Boissy, A. (1995). Fear and fearfulness in animals. *Q. Rev. Biol.* 70(2), 165–191.

Boissy, A., Aubert, A., Désiré, L., Greiveldinger, L., Delval, E. and Veisser, I. (2011). Cognitive sciences to relate ear postures to emotions in sheep. *Anim. Welf.* 20, 47–56.

Boissy, A., Manteuffel, G., Jensen, M. B., Oppermann Moe, R. O., Spruijt, B., Keeling, L. J., Winckler, C., Forkman, B., Dimitrov, I., Langbein, J., Bakken, M., Veissier, I. and Aubert, A. (2007). Assessment of positive emotions in animals to improve their welfare. *Physiol. Behav.* 92(3), 375–397.

Boivin, X., Le Neindre, P., Garel, J. P. and Chupin, J. M. (1994). Influence of breed and rearing management on cattle reactions during human handling. *Appl. Anim. Behav. Sci.* 39(2), 115–122.

Bouissou, M.-F. (1980). Social relationships in domestic cattle under modern management techniques. *Ital. J. Zool. (Modena)* 47(3–4), 343–353.

Bouissou, M.-F., Boissy, A., Le Neindre, P. and Veissier, I. (2001). The social behaviour of cattle. In: *Social Behaviour in Farm Animals* Keeling, L. J. and Gonyou, H. W. (Eds). CABI Publishing, Oxon, UK, pp. 113–145.

Bourguet, C., Deiss, V., Boissy, A. and Terlouw, E. M. C. (2015). Young Blond d'Aquitaine, Angus and Limousin bulls differ in emotional reactivity: relationships with animal traits, stress reactions at slaughter and post-mortem muscle metabolism. *Appl. Anim. Behav. Sci.* 164, 41–55.

Breuer, K., Hemsworth, P. H., Barnett, J. L., Matthews, L. R. and Coleman, G. J. (2000). Behavioural response to humans and the productivity of commercial dairy cows. *Appl. Anim. Behav. Sci.* 66(4), 273-288.

Burrow, H. M., Seifert, G. W. and Corbet, N. J. (1988). A new technique for measuring temperament in cattle. *Proc. Aust. Soc. Anim. Prod.* 17, 154-157.

BVA/BCVA. (2017). Joint position statement. Analgesia in Calves. Available at: www.bva.co.uk/uploadedFiles/Content/News,_campaigns_and_policies/Policies/Ethics_and_welfare/Analgesia%20in%20calves%20BVA%20branded.pdf.

Byrd, C. J., Craig, B. A., Eicher, S. D., Radcliffe, J. S. and Lay, D. C. (2019). Short communication: assessment of disbudding pain in dairy calves using nonlinear measures of heart rate variability. *J. Dairy Sci.* 102(9), 8410-8416.

Casoni, D., Mirra, A., Suter, M. R., Gutzwiller, A. and Spadavecchia, C. (2019). Can disbudding of calves (one versus four weeks of age) induce chronic pain? *Physiol. Behav.* 199, 47-55.

Chebel, R. C., Silva, P. R. B., Endres, M. I., Ballou, M. A. and Luchterhand, K. L. (2016). Social stressors and their effects on immunity and health of periparturient dairy cows. *J. Dairy Sci.* 99(4), 3217-3228.

Chen, J. M., Schütz, K. E. and Tucker, C. B. (2013). Dairy cows use and prefer feed bunks fitted with sprinklers. *J. Dairy Sci.* 96(8), 5035-5045.

Collings, L. K. M., Weary, D. M., Chapinal, N. and von Keyserlingk, M. A. G. (2011). Temporal feed restriction and overstocking increase competition for feed by dairy cattle. *J. Dairy Sci.* 94(11), 5480-5486.

Cook, N. B., Mentink, R. L., Bennett, T. B. and Burgi, K. (2007). The effect of heat stress and lameness on time budgets of lactating dairy cows. *J. Dairy Sci.* 90(4), 1674-1682.

Curley, K. O., Jr., Paschal, J. C., Welsh, T. H., Jr. and Randel, R. D. (2006a). Technical note: exit velocity as a measure of cattle temperament is repeatable and associated with serum concentration of cortisol in Brahman bulls. *J. Anim. Sci.* 84(11), 3100-3103.

Curley, K. O., Jr., Schuehle Pfeiffer, C. E., King, D. A., Savell, J. W., Vann, R. C., Welsh, T. H., Jr. and Randel, R. D. (2006b). Relationships of cattle temperament and physiologic responses to handling during typical management situations. *J. Anim. Sci.* 84 (Suppl. 2), 32.

Darwin, C. (1872). *The Expression of Emotions in Animals and Man*. John Murray, London.

Dawkins, M. S. (1990). From an animal's point of view: motivation, fitness and animal welfare. *Behav. Brain Sci.* 13(1), 1-9.

Dawkins, M. S. (2005). The science of suffering. In: *Mental Health and Well-Being in Animals* McMillan, F. D. (Ed.). Blackwell Publishing Professional, Iowa, USA, pp. 47-55.

Dawkins, M. S. (2006). Through animal eyes: what behaviour tells us. *Appl. Anim. Behav. Sci.* 100(1-2), 4-10.

D'Eath, R. B., Tolkamp, B. J., Ilias Kyriazakis, I. and Lawrence, A. B. (2009). 'Freedom from hunger' and preventing obesity: the animal welfare implications of reducing food quantity or quality. *Anim. Behav.* 77(2), 275-288.

de Boyer des Roches, A., Lussert, A., Faure, M., Herry, V., Rainard, P., Durand, D., Wemelsfelder, F. and Foucras, G. (2018). Dairy cows under experimentally-induced Escherichia coli mastitis show negative emotional states assessed through qualitative behaviour assessment. *Appl. Anim. Behav. Sci.* 206, 1-11.

Désiré, L., Veissier, I., Després, G. and Boissy, A. (2004). On the way to assess emotions in animals: do lambs (Ovis aries) evaluate an event through its suddenness, novelty, or unpredictability? *J. Comp. Psych.* 118(4), 363-374.

Désiré, L., Veissier, I., Després, G., Delval, E., Toporenko, G. and Boissy, A. (2006). Appraisal process in sheep (Ovis aries): interactive effect of suddenness and unfamiliarity on cardiac and behavioral responses. *J. Comp. Psych.* 120(3), 280–287.

DeVries, T. J., von Keyserlingk, M. A. G. and Weary, D. M. (2004). Effect of feeding space on the inter-cow distance, aggression, and feeding behavior of free-stall housed lactating dairy cows. *J. Dairy Sci.* 87(5), 1432–1438.

Dhabhar, F. S. and McEwen, B. S. (2001). Bidirectional effects of stress and glucocorticoid hormones on immune function: possible explanations for paradoxical observations. In: *Psychoneuroimmunology* (3rd edn., Vol. 1) Ader, R., Felten, D. L. and Cohen, N. (Eds). Academic Press, London, pp. 301–338.

Duncan, I. J. H. (1992). Measuring preferences and the strength of preferences. *Poult. Sci.* 71(4), 658–663.

Ede, T., Lecorps, B., von Keyserlingk, M. A. G. and Weary, D. M. (2019a). Calf aversion to hot-iron disbudding. *Sci. Rep.* 9(1), 5344.

Ede, T., von Keyserlingk, M. A. G. and Weary, D. M. (2018). Approach-aversion in calves following injections. *Sci. Rep.* 8(1), 9443.

Ede, T., von Keyserlingk, M. A. G. and Weary, D. M. (2019b). Assessing the affective component of pain, and the efficacy of pain control, using conditioned place aversion in calves. *Biol. Lett.* 15(10), 20190642.

Ekman, P. and Friesen, W. (1978). *Facial Action Coding System: A Technique for the Measurement of Facial Movement*. Consulting Psychologists Press, Palo Alto.

Elsasser, T . H., Klasing, K. C., Filipov, N. and Thompson, F. (2000). The metabolic consequences of stress: targets for stress and priorities of nutrient use. In: *Biology of Animal Stress: Implications for Animal Welfare*, Moberg, G. P. and Mench, J. A. (Eds). CAB International, Wallingford, Oxon, UK, pp. 77–110.

Faulkner, P. M. and Weary, D. M. (2000). Reducing pain after dehorning in dairy calves. *J. Dairy Sci.* 83(9), 2037–2041.

FAWC (2019). *Opinion on the Welfare of Cattle Kept for Beef Production*. Farm Animal Welfare Committee, London.

Fell, L. R., Colditz, I. G., Walker, K. H. and Watson, D. L. (1999). Associations between temperament, performance and immune function in cattle entering a commercial feedlot. *Aust. J. Exp. Agric.* 39(7), 795–802.

Fitzpatrick, C. E., Chapinal, N., Petersson-Wolfe, C. S., DeVries, T. J., Kelton, D. F., Duffield, T. F. and Leslie, K. E. (2013). The effect of meloxicam on pain sensitivity, rumination time, and clinical signs in dairy cows with endotoxin-induced clinical mastitis. *J. Dairy Sci.* 96(5), 2847–2856.

Fordyce, G., Goddard, M. E., Tyler, R., Williams, G. and Toleman, M. A. (1985). Temperament and bruising of Bos indicus cross cattle. *Aust. J. Exp. Agric.* 25(2), 282–288.

Forkman, B., Boissy, A., Meunier-Salaun, M. C., Canalie, E. and Jones, R. B. (2007). A critical review of fear tests used on cattle, pigs, sheep, poultry and horses. *Physiol. Behav.* 92(3), 340–374.

Franchi, G. A., Herskin, M. S. and Jensen, M. B. (2019). Dairy cows fed a low energy diet before dry-off show signs of hunger despite ad libitum access. *Sci. Rep.* 9(1), 16159.

Fregonesi, J. A., Tucker, C. B. and Weary, D. M. (2007). Overstocking reduces lying time in dairy cows. *J. Dairy Sci.* 90(7), 3349–3354.

Frondelius, L., Järvenranta, K., Koponen, T. and Mononen, J. (2015). The effects of body posture and temperament on heart rate variability in dairy cows. *Physiol. Behav.* 139, 437–441.

Fuchs, P. N. and McNabb, C. T. (2012). The place escape/avoidance paradigm: a novel method to assess nociceptive processing. *J. Integr. Neurosci.* 11(1), 61-72.

Fujiwara, M., Haskell, M., Macrae, A. and Rutherford, K. (2019). Effects of stocking density during the dry period on dairy cow physiology, metabolism and behaviour. *J. Dairy Res.* 86(3), 283-290.

Fujiwara, M., Haskell, M. J., Macrae, A. I. and Rutherford, K. M. D. (2018). Survey of dry cow management on UK commercial dairy farms. *Vet. Rec.* 183(9), 297.

Gabriel, A. F., Paoletti, G., Della Seta, D., Panelli, R., Marcus, M. A. E., Farabollini, F., Carli, G. and Joosten, E. A. J. (2010). Enriched environment and the recovery from inflammatory pain: social versus physical aspects and their interaction. *Behav. Brain Res.* 208(1), 90-95.

Gibbons, J., Lawrence, A. B. and Haskell, M. J. (2009). Responsiveness of dairy cows to human approach and novel stimuli. *Appl. Anim. Behav. Sci.* 116, 163-173.

Gibson, T. J., Johnson, C. B., Stafford, K. J., Mitchinson, S. L. and Mellor, D. J. (2007). Validation of the acute electroencephalographic responses of calves to noxious stimulus with scoop dehorning. *N. Z. Vet. J.* 55(4), 152-157.

Gladden, N., Ellis, K., Martin, J., Viora, L. and McKeegan, D. (2019). A single dose of ketoprofen in the immediate postpartum period has the potential to improve dairy calf welfare in the first 48h of life. *Appl. Anim. Behav. Sci.* 212, 19-29.

Gleerup, K. B., Andersen, P. H., Munksgaard, L. and Forkman, B. (2015). Pain evaluation in dairy cattle. *Appl. Anim. Behav. Sci.* 171, 25-32.

Goma, A. A., Pearce, G. P., Uddin, J., Rimon, E., Davies, H. and Phillips, C. J. C. (2018). A forced lateralisation test for dairy cows and its relation to their behaviour. *Appl. Anim. Behav. Sci.* 207, 8-19.

Grandin, T. (1993). Behavioral agitation during handling of cattle is persistent over time. *Appl. Anim. Behav. Sci.* 36, 1-9.

Gray, J. A. (1987). *The Psychology of Fear and Stress*. Cambridge University Press, Cambridge, UK.

Greter, A. M., Miller-Cushon, E. K., McBride, B. W., Widowski, T. M., Duffield, T. F. and DeVries, T. J. (2015). Short communication: limit feeding affects behavior patterns and feeding motivation of dairy heifers. *J. Dairy Sci.* 98(2), 1248-1254.

Hahn, G. L. (1999). Dynamic responses of cattle to thermal heat loads. *J. Anim. Sci.* 77 (Suppl. 2), 10-20.

Haskell, M. J., Masłowska, K., Bell, D. J., Roberts, D. J. and Langford, F. M. (2013). The effect of a view to the surroundings and microclimate variables on use of a loafing area. *Appl. Anim. Behav. Sci.* 147, 28-33.

Herskin, M. S. and Nielsen, B. H. (2018). Welfare effects of the use of a combination of local anesthesia and NSAID for disbudding analgesia in dairy calves–reviewed across different welfare concerns. *Front. Vet. Sci.* 5, 117.

Hoffman, P. C., Simson, C. R. and Wattiaux, M. (2007). Limit feeding of gravid Holstein heifers: effect on growth, manure nutrient excretion, and subsequent early lactation performance. *J. Dairy Sci.* 90(2), 946-954.

Hokkanen, A. H., Wikman, I., Korhonen, T., Pastell, M., Valros, A., Vainio, O. and Hänninen, L. (2015). Perceptions and practices of Finnish dairy producers on disbudding pain in calves. *J. Dairy Sci.* 98(2), 823-831.

Hopster, H., Bruckmaier, R. M., Van der Werf, J. T. N., Korte, S. M., Macuhova, J., Korte-Bouws, G. and van Reenen, C. G. (2002). Stress responses during milking; comparing conventional and automatic milking in primiparous dairy cows. *J. Dairy Sci.* 85(12), 3206–3216.

Hughes, B. O. and Black, A. J. (1973). The preference of domestic hens for different types of battery cage floor. *Br. Poult. Sci.* 14(6), 615–619.

Huxley, J. N. and Whay, H. R. (2006). Current attitudes of cattle practitioners to pain and the use of analgesics in cattle. *Vet. Rec.* 159(20), 662–668.

Huzzey, J. M., DeVries, T. J., Valois, P. and von Keyserlingk, M. A. G. (2006). Stocking density and feed barrier design affect the feeding and social behavior of dairy cattle. *J. Dairy Sci.* 89(1), 126–133.

Jensen, M. B., Pedersen, L. J. and Munksgaard, L. (2005). The effect of reward duration on demand functions for rest in dairy heifers and lying requirements as measured by demand functions. *Appl. Anim. Behav. Sci.* 90(3–4), 207–217.

Jimenez, R. E., Adcock, S. J. J. and Tucker, C. B. (2019). Acute pain responses in dairy calves undergoing cornual nerve blocks with or without topical anaesthetic. *J. Dairy Sci.* 102(4), 3431–3438.

Kadzere, C. T., Murphy, M. R., Silanikove, N. and Maltz, E. (2002). Heat stress in lactating dairy cows: a review. *Livest. Prod. Sci.* 77(1), 59–91.

Kappel, S., Mendl, M. T., Barrett, D. C., Murrell, J. C. and Whay, H. R. (2017). Lateralized behaviour as indicator of affective state in dairy cows. *PLoS ONE* 12(9), e0184933.

Kilgour, R. J., Melville, G. J. and Greenwood, P. L. (2006). Individual differences in the reaction of beef cattle to situations involving social isolation, close proximity of humans, restraint and novelty. *Appl. Anim. Behav. Sci.* 99(1–2), 21–40.

King, D. A., Schuehle Pfeiffer, C. E., Randel, R. D., Welsh, T. H., Jr., Oliphint, R. A., Baird, B. E., Curley, K. O., Jr., Vann, R. C., Hale, D. S. and Savell, J. W. (2006). Influence of animal temperament and stress responsiveness on the carcass quality and beef tenderness of feedlot cattle. *Meat Sci.* 74(3), 546–556.

Kingma, B. R. M., Frijns, A. J. H., Schellen, L. and van Marken Lichtenbelt, W. D. (2014). Beyond the classic thermoneutral zone Including thermal comfort. *Temperature* 1(2), 142–149.

Kirkden, R. D. and Pajor, E. A. (2006). Using preference, motivation and aversion tests to ask scientific questions about animals' feelings. *Appl. Anim. Behav. Sci.* 100(1–2), 29–47.

Kleinhenz, M. D., Van Engen, N. K., Gorden, P. J., Ji, J., Walsh, P. and Coetzee, J. F. (2017). Effects of transdermal flunixin meglumine on pain biomarkers at dehorning in calves. *J. Anim. Sci.* 95(5), 1993–2000.

Koolhaas, J. M., Bartolomucci, A., Buwalda, B., de Boer, S. F., Flugge, G., Korte, S. M., Meerlo, P., Murison, R., Olivier, B., Palanza, P., Richter-Levin, G., Sgoifo, A., Steimer, T., Stiedl, O., van Dijk, G., Wohr, M. and Fuchs, E. (2011). Stress revisited: a critical evaluation of the stress concept. *Neurosci. Biobehav. Rev.* 35(5), 1291–1301.

Kovács, L., Kézér, F. L., Jurkovich, V., Kulcsár-Huszenicza, M. and Tőzsér, J. (2015a). Heart rate variability as an indicator of chronic stress caused by lameness in dairy cows. *PLoS ONE* 10(8), e0134792.

Kovács, L. L., Kézér, F., Tőzsér, J., Szenci, O., Péter Póti, P. and Pajor, F. (2015b). Heart rate and heart rate variability in dairy cows with different temperament and behavioural reactivity to humans. *PLoS ONE* 10, e0136294.

Krawczel, P. D., Klaiber, L. B., Butzler, R. E., Klaiber, L. M., Dann, H. M., Mooney, C. S. and Grant, R. J. (2012). Short-term increases in stocking density affect the lying and social behavior, but not the productivity, of lactating Holstein dairy cows. *J. Dairy Sci.* 95(8), 4298–4308.

Lago, A., McGuirk, S. M., Bennett, T. B., Cook, N. B. and Nordlund, K. V. (2006). Calf respiratory disease and pen microenvironments in naturally ventilated calf barns in winter. *J. Dairy Sci.* 89(10), 4014–4025.

Langford, D. J., Bailey, A. L., Chanda, M. L., Clarke, S. E., Drummond, T. E., Echols, S., Glick, S., Ingrao, J., Klassen-Ross, T., LaCroix-Fralish, M. L., Matsumiya, L., Sorge, R. E., Sotocinal, S. G. G., Tabaka, J. M., Wong, D., van den Maagdenberg, A. M. J. M., Ferrari, M. D., Craig, K. D. and Mogil, J. S. (2010). Coding of facial expressions of pain in the laboratory mouse. *Nat. Methods* 7(6), 447–449. doi:10.1038/nmeth.1455.

Langford, F. M., Rutherford, K. M. D., Sherwood, L., Jack, M. C., Lawrence, A. B. and Haskell, M. J. (2011). Behavior of cows during and after peak feeding time on organic and conventional dairy farms in the UK. *J. Dairy Sci.* 94(2), 746–753.

Laven, R., Chambers, P. and Stafford, K. (2012). Using non-steroidal anti-inflammatory drugs around calving: maximising comfort, productivity and fertility. *Vet. J.* 192(1), 8–12.

Le Neindre, P., Trillat, G., Sapa, J., Menissier, F., Bonnet, J. N. and Chupin, J. M. (1995). Individual differences in docility in Limousin cattle. *J. Anim. Sci.* 73(8), 2249–2253.

Lecorps, B., Ludwig, B. R., von Keyserlingk, M. A. G. and Weary, D. M. (2019). Pain-induced pessimism and anhedonia: evidence from a novel probability-based judgment bias test. *Front. Behav. Neurosci.* 13, 54.

Lecorps, B., Weary, D. M., Weary, D. M. and Von Keyserlingk, M. A. G. (2018). Pessimism and fearfulness in dairy calves. *Sci. Rep.* 8, 1421.

Lee, C., Cafe, L. M., Robinson, S., Doyle, R. E., Lea, J., Small, A. H., Colditz, I. G. (2018). Anxiety influences attention bias but not flight speed and crush score in beef cattle. *Appl. Anim. Behav. Sci.* 205, 210–215.

Legrand, A., Schütz, K. E. and Tucker, C. B. (2011). Using water to cool cattle: behavioral and physiological changes associated with voluntary use of cow showers. *J. Dairy Sci.* 94(7), 3376–3386.

Li, Z., Lu, Y. F., Li, C. L., Wang, Y., Sun, W., He, T., Chen, X. F., Wang, X. L. and Chen, J. (2014). Social interaction with a cagemate in pain facilitates subsequent spinal nociception via activation of the medial prefrontal cortex in rats. *Pain* 155(7), 1253–1261.

Mandel, R., Harazy, H., Gygax, L., Nicol, C. J., Ben-David, A., Whay, H. R. and Klement, E. (2018). Short communication: detection of lameness in dairy cows using a grooming device. *J. Dairy Sci.* 101(2), 1511–1517.

Matthews, L. R. and Ladewig, J. (1994). Environmental requirements of pigs measured by behavioural demand functions. *Anim. Behav.* 47(3), 713–719.

McDonald, P. V., von Keyserlingk, M. A. G. and Weary, D. M. (2020). Hot weather increases competition between dairy cows at the drinker. *J. Dairy Sci.* 103(4), 3447–3458.

McEwen, B. S. (2005). Stressed or stressed out: what is the difference? *J. Psychiatry Neurosci.* 30(5), 315–318.

McEwan, B. S. and Stellar, E. (1993). Stress and the individual: mechanisms leading to disease. *Arch. Intern. Med.* 153(18), 2093–2101.

McEwen, B. S. and Wingfield, J. C. (2003). The concept of allostasis in biology and biomedicine. *Horm. Behav.* 43(1), 2–15.

Mendl, M. and Paul, E. S. (2020). Animal affect and decision-making. *Neurosci. Biobehav. Rev.* 112, 144–163.

Metz, J. H. M. (1985). The reaction of cows to a short-term deprivation of lying. *Appl. Anim. Behav. Sci.* 13(4), 301–307.

Miller-Cushon, E. K., Horvath, K. C., Fabris, T. F., Dahl, G. E. and Laporta, J. (2019). Short communication: effects of mammary biopsy in the dry period on activity and feeding behavior of dairy cows. *J. Dairy Sci.* 102(12), 11453–11458.

Mintline, E. M., Stewart, M., Rogers, A. R., Cox, N. R., Verkerk, G. A., Stookey, J. M., Webster, J. R. and Tucker, C. B. (2013). Play behavior as an indicator of animal welfare: disbudding in dairy calves. *Appl. Anim. Behav. Sci.* 144(1–2), 22–30.

Mir, R. A., Kleinhenz, M. D., Coetzee, J. F., Allen, H. K. and Kudva, I. T. (2019). Fecal microbiota changes associated with dehorning and castration stress primarily affects light-weight dairy calves. *PLoS ONE* 14(1), e0210203.

Moberg, G. P. (2000). Biological response to stress: implications for animal welfare. In: *The Biology of Animal Stress: Basic Principles and Implications for Animal Welfare*, Moberg, G. P. and Mench, J. A. (Eds). CAB International Publishing, Wallingford, pp. 1–21.

Moggy, M. A., Pajor, E. A., Thurston, W. E., Parker, S., Greter, A. M., Schwartzkopf-Genswein, K. S. and Windeyer, M. C. (2017). Management practices associated with pain in cattle on western Canadian cow-calf operations: a mixed methods study. *J. Anim. Sci.* 95, 958–969.

Mogil, J. S. (2009). Animal models of pain: progress and challenges. *Nat. Rev. Neurosci.* 10(4), 283–294.

Monk, J. E., Doyle, R. E., Colditz, I. G., Belson, S., Cronin, G. M. and Lee, C. (2018). Towards a more practical attention bias test to assess affective state in sheep. *PLoS ONE* 13(1), e0190404.

Morgan, C. A., McIlvaney, K., Dwyer, C. M. and Lawrence, A. B. (2009). Assessing the welfare challenges to out-wintered pregnant suckler cows. *Animal* 3(8), 1167–1174.

Neave, H. W., Daros, R. R., Costa, J. H. C., von Keyserkingk, M. A. G. and Weary, D. M. (2013). Pain and pessimism: dairy calves exhibit negative judgment bias following hot-iron disbudding. *PLoS ONE* 8(12), e80556.

Newby, N. C., Tucker, C. B., Pearl, D. L., LeBlanc, S. J., Leslie, K. E., von Keyserlingk, M. A. G. and Duffield, T. F. (2014). An investigation of the effects of ketoprofen following rumen fistulation surgery in lactating dairy cows. *Can. Vet. J.* 55(5), 442–448.

Nielsen, B. L. (2020). *Asking Animals: An Introduction to Animal Testing*. CABI Publishing, Oxfordshire, UK.

Odensten, M. O., Holtenius, K. and Persson Waller, K. P. (2007). Effects of two different feeding strategies During dry-off on certain health aspects of dairy cows. *J. Dairy Sci.* 90(2), 898–907.

O'Leary, N. W., Byrne, D. T., O'Connor, A. H. and Shalloo, L. (2020). Invited review: cattle lameness detection with accelerometers. *J. Dairy Sci.* 103(5), 3895–3911.

Ominski, K. H., Kennedy, A. D., Wittenberg, K. M. and Moshtaghi Nia, S. A . (2002). Physiological and production responses to feeding schedule in lactating dairy cows exposed to short-term, moderate heat stress. *J. Dairy Sci.* 85(4), 730–737.

Pajor, E. A., Rushen, J. and de Passillé, A. M. B. (2003). Dairy cattle's choice of handling treatments in a Y-maze. *Appl. Anim. Behav. Sci.* 80(2), 93–107.

Palme, R. (2019). Non-invasive measurement of glucocorticoids: advances and problems. *Physiol. Behav.* 199, 229–243.

Paul, E. S., Harding, E. J. and Mendl, M. (2005). Measuring emotional processes in animals: the utility of a cognitive approach. *Neurosci. Biobehav. Rev.* 29(3), 469–491.

Phillips, C. J. C. and Rind, M. I. (2001). The effects on production and behavior of mixing uniparous and multiparous cows. *J. Dairy Sci.* 84(11), 2424–2429.

Phillips, C. J. C., Oevermans, H., Syrett, K. L., Jespersen, A. Y. and Pearce, G. P. (2015). Lateralization of behavior in dairy cows in response to conspecifics and novel persons. *J. Dairy Sci.* 98(4), 2389–2400.

Polsky, L. and von Keyserlingk, M. A. G. (2017). Invited review: effects of heat stress on dairy cattle welfare. *J. Dairy Sci.* 100(11), 8645–8657.

Price, E. O. (1984). Behavioral aspects of animal domestication. *Q. Rev. Biol.* 59(1), 1–32.

Proudfoot, K. L., Veira, D. M., Weary, D. M. and von Keyserlingk, M. A. G. (2009). Competition at the feed bunk changes the feeding, standing, and social behavior of transition dairy cows. *J. Dairy Sci.* 92(7), 3116–3123.

Raja, S. N., Carr, D. B., Cohen, M., Finnerup, N. B., Flor, H., Gibson, S., Keefe, F. J., Mogil, J. S., Ringkamp, M., Sluka, K. A., Song, X. J., Stevens, B., Sullivan, M. D., Tutelman, P. R., Ushida, T. and Vader, K. (2020). The revised International Association for the Study of Pain definition of pain: concepts, challenges, and compromises. Pain 161(9), 1976–1982.

Rasmussen, D. B., Fogsgaard, K., Røntved, C. M., Klaas, I. C. and Herskin, M. S. (2011). Changes in thermal nociceptive responses in dairy cows following experimentally induced Escherichia coli mastitis. *Acta. Vet. Scand.* 53, 32.

Reefman, N., Wechsler, B. and Gygax, L. (2009). Behavioural and physiological assessment of positive and negative emotion in sheep. *Anim. Behav.* 78(3), 651–659.

Remnant, J. G., Tremlett, A., Huxley, J. N. and Hudson, C. D. (2017). Clinician attitudes to pain and use of analgesia in cattle: where are we 10 years on? *Vet. Rec.* 181(15), 400.

Rioja-Lang, F. C., Roberts, D. J., Healy, S. D., Lawrence, A. B. and Haskell, M. J. (2009). Dairy cows trade-off feed quality with proximity to a dominant individual in Y-maze choice tests. *Appl. Anim. Behav. Sci.* 117(3–4), 159–164.

Rioja-Lang, F. C., Roberts, D. J., Healy, S. D., Lawrence, A. B. and Haskell, M. J. (2012). Dairy cow feeding space requirements assessed in a Y-maze choice test. *J. Dairy Sci.* 95(7), 3954–3960.

Robins, A. and Phillips, C. J. C. (2010). Lateralised visual processing in domestic cattle herds responding to novel and familiar stimuli. *Laterality* 15(5), 514–534.

Rogers, L. J. (2002). Lateralization in vertebrates: its early evolution, general pattern, and development. *Adv. Study Behav.* 31, 107–161.

Rogers, L. J. (2010). Relevance of brain and behavioural lateralization to animal welfare. *Appl. Anim. Behav. Sci.* 127(1–2), 1–11.

Rushen, J., de Passille, A. M. and Munksgaard, L. (1999). Fear of people by cows and effects on milk yield, behavior, and heart rate at milking. *J. Dairy Sci.* 82(4), 720–727.

Rutherford, K. M. D., Donald, R. D., Lawrence, A. B. and Wemelsfelder, F. (2012). Qualitative behavioural assessment of emotionality in pigs. *Appl. Anim. Behav. Sci.* 139(3–4), 218–224.

Rutherford, K. M. D., Langford, F. M., Jack, M. C., Sherwood, L., Lawrence, A. B. and Haskell, M. J. (2009). Lameness prevalence and risk factors in organic and non-organic dairy herds in the United Kingdom. *Vet. J.* 180(1), 95–105.

Sandem, A. I., Braastad, B. O. and Bøe, K. E. (2002). Eye white may indicate emotional state on a frustration-contentedness axis in dairy cows. *Appl. Anim. Behav. Sci.* 79(1), 1–10.

Schaefer, A. L., Matthews, L. R., Cook, N. J., Webster, J. and Scott, S. L. (2002). Novel non-invasive measures of animal welfare. *Animal Welfare and Behaviour: From Science to Solution, Joint NAWAC/ISAE Conference*. Hamilton, New Zealand, 27–28 June 2002.

Schütz, K. E., Davidson, D. and Matthews, L. (2006). Do different levels of moderate feed deprivation in dairy cows affect feeding motivation? *Appl. Anim. Behav. Sci.* 101(3-4), 253-263.

Schütz, K. E., Cox, N. R. and Matthews, L. R. (2008). How important is shade to dairy cattle? Choice between shade or lying following different levels of lying deprivation. *Appl. Anim. Behav. Sci.* 114(3-4), 307-318.

Schütz, K. E., Rogers, A. R., Poulouin, Y. A., Cox, N. R. and Tucker, C. B. (2010). The amount of shade influences the behavior and physiology of dairy cattle. *J. Dairy Sci.* 93(1), 125-133.

Schütz, K. E., Rogers, A. R., Cox, N. R., Webster, J. R. and Tucker, C. B. (2011). Dairy cattle prefer shade over sprinklers: effects on behavior and physiology. *J. Dairy Sci.* 94(1), 273-283.

Shaffer, F., McCraty, R. and Zerr, C. L. (2014). A healthy heart is not a metronome: an integrative review of the heart's anatomy and heart rate variability. *Front. Psychol.* 5, 1040. doi:10.3389/fpsyg.2014.01040.

Silanikove, N. (2000). Effects of heat stress on the welfare of extensively managed domestic ruminants. *Livest. Prod. Sci.* 67(1-2), 1-18.

Solano, L., Barkema, H. W., Pajor, E. A., Mason, S., LeBlanc, S. J., Zaffino Heyerhoff, J. C., Nash, C. G. R., Haley, D. B., Vasseur, E., Pellerin, D., Rushen, J., de Passillé, A. M . and Orsel, K. (2015). Prevalence of lameness and associated risk factors in Canadian Holstein-Friesian cows housed in freestall barns. *J. Dairy Sci.* 98(10), 6978-6991.

Sotocinal, S. G., Sorge, R. E., Zaloum, A., Tuttle, A. H., Martin, L. J., Wieskopf, J. S., Mapplebeck, J. C. S., Wei, P., Zhan, S., Zhang, S., McDougall, J. J., King, O. D. and Mogil, J. S. (2011). The Rat Grimace Scale: a partially automated method for quantifying pain in the laboratory rat via facial expressions. *Mol. Pain* 7, 55.

Stewart, M., Stafford, K. J., Dowling, S. K., Schaefer, A. L. and Webster, J. R. (2008). Eye temperature and heart rate variability of calves disbudded with or without local anaesthetic. *Physiol. Behav.* 93(4-5), 789-797.

Stewart, M., Webster, J. R., Verkerk, G. A., Colyn, J. J. and Schaefer, A. L. (2005a). Infrared thermography as a non-invasive measure of stress in dairy cows. *J. Anim. Sci.* 83(Suppl. 1), 374.

Stewart, M., Webster, J. R., Schaefer, A. L., Cook, N. J. and Scott, S. L. (2005b). Infrared thermography as a non-invasive tool to study animal welfare. *Anim. Welf.* 14, 319-325.

Sutherland, M. A., Mellor, D. J., Stafford, K. J., Gregory, N. G., Bruce, R. A. and Ward, R. N. (2002). Cortisol responses to dehorning of calves given a 5-h local anaesthetic regimen plus phenylbutazone, ketoprofen, or adrenocorticotropic hormone prior to dehorning. *Res. Vet. Sci.* 73(2), 115-123.

Thompson, A. J., Weary, D. M., Bran, J. A., Daros, R. R., Hötzel, M. J. and von Keyserlingk, M. A. G. (2019). Lameness and lying behavior in grazing dairy cows. *J. Dairy Sci.* 102(7), 6373-6382.

Toates, F. (1986). *Motivational Systems. Problems in the Behavioural Sciences*. Cambridge University Press, Cambridge.

Tschoner, T., Behrendt-Wippermann, M., Rieger, A., Metzner, M., Knubben-Schweizer, G., Reichmann, F. and Feist, M. (2018). Course of plasma substance P concentrations during umbilical surgery in calves. *Berl. Munch. Tierarztl Wochenschr.* 131, 522-528.

Tucker, C. B., Lacy-Hulbert, S. J. and Webster, J. R. (2009). Effect of milking frequency and feeding level before and after dry off on dairy cattle behavior and udder characteristics. *J. Dairy Sci.* 92(7), 3194-3203.

Tucker, C. B., Rogers, A. R. and Schütz, K. E. (2008). Effect of solar radiation on dairy cattle behaviour, use of shade and body temperature in a pasture-based system. *Appl. Anim. Behav. Sci.* 109(2-4), 141-154.

Tucker, C. B., Rogers, A. R., Verkerk, G. A., Kendall, P. E., Webster, J. R. and Matthews, L. R. (2007). Effects of shelter and body condition on the behaviour and physiology of dairy cattle in winter. *Appl. Anim. Behav. Sci.* 105(1-3), 1-13.

Valizaheh, R., Veira, D. M. and von Keyserlingk, M. A. G. (2008). Behavioural responses by dairy cows provided two hays of contrasting quality at dry-off. *Appl. Anim. Behav. Sci.* 109(2-4), 190-200.

Val-Laillet, D., Guesdon, V., von Keyserlingk, M. A. G., de Passillé, A. M. and Rushen, J. (2009). Allogrooming in cattle: relationships between social preferences, feeding displacements and social dominance. *Appl. Anim. Behav. Sci.* 116(2-4), 141-149.

Van laer, E., Ampe, B., Moons, C., Sonck, B. and Tuyttens, F. A. M. (2015). Wintertime use of natural versus artificial shelter by cattle in nature reserves in temperate areas. *Appl. Anim. Behav. Sci.* 163, 39-49.

Van Reenan, C. G., O'Connell, N. E., van der Werf, J. T. N., Korte, S. M., Hopster, H., Jones, R. B. and Blokhuis, H. J. (2005). Responses of calves to acute stress: individual consistency and relations between behavioral and physiological measures. *Physiol. Behav.* 85(5), 557-570.

Vindevoghel, T. V., Fleming, P. A., Hyndman, T. H., Musk, G. C., Laurence, M. and Collins, T. (2019). Qualitative behavioural assessment of Bos indicus cattle after surgical castration. *Appl. Anim. Behav. Sci.* 211, 95-102.

Vizzotto, E. F., Fischer, V., Thaler Neto, A., Abreu, A. S., A. S., Stumpt, M. T., Werncke, D., Schmidt, F. A. and McManus, C. M. (2015). Access to shade changes behavioral and physiological attributes of dairy cows during the hot season in the subtropics. *Animal* 9(9), 1559-1566.

von Borell, E., Langbein, J., Després, G., Hansen, S., Leterrier, C., Marchant-Forde, J., Marchant-Forde, R., Minero, M., Mohr, E., Prunier, A., Valance, D. and Veissier, I. (2007). Heart rate variability as a measure of autonomic regulation of cardiac activity for assessing stress and welfare in farm animals: a review. *Physiol. Behav.* 92(3), 293-316.

Waiblinger, S., Boivin, X., Pedersen, V., Tosi, M.-V., Janczak, A. M., Visser, E. K., Jones, R. B. (2006). Assessing the human-animal relationship in farmed species: a critical review. *Appl. Anim. Behav. Sci.* 101(3-4), 185-242.

Waiblinger, S., Menke, C. and Fölsch, D. W. (2003). Influences on the avoidance and approach behaviour of dairy cows towards humans on 35 farms. *Appl. Anim. Behav. Sci.* 84(1), 23-39.

Waller, B. M. and Micheletta, J. (2013). Facial expression in nonhuman animals. *Emot. Rev.* 5(1), 54-59.

Watts, J. M. and Stookey, J. M. (2000). Vocal behaviour in cattle: the animal's commentary on its biological processes and welfare. *Appl. Anim. Behav. Sci.* 67(1-2), 15-33.

Weary, D. M., Droege, P. and Braithwaite, V. A. (2017). Behavioral evidence of felt emotions. *Adv. Study Behav.* 49, 27-48.

Webster, J. R., Stewart, M., Rogers, A. R. and Verkerk, G. A. (2008). Assessment of welfare from physiological and behavioural responses of New Zealand dairy cows exposed to cold and wet conditions. *Anim. Welf.* 17, 19-26.

Weigele, H. C., Gygax, L., Steiner, A., Wechsler, B. and Burla, J. B. (2018). Moderate lameness leads to marked behavioral changes in dairy cows. *J. Dairy Sci.* 101(3), 2370-2382.

Wemelsfelder, F. (2007). How animals communicate quality of life: the qualitative assessment of behaviour. *Anim. Welf.* 16, 25–31.

Wemelsfelder, F., Hunter, E. A., Mendl, M. T. and Lawrence, A. B. (2000). The spontaneous qualitative assessment of behavioural expressions in pigs: first explorations of a novel methodology for integrative animal welfare measurement. *Appl. Anim. Behav. Sci.* 67(3), 193–215.

West, J. W. (2003). Effects of heat-stress on production in dairy cattle. *J. Dairy Sci.* 86(6), 2131–2144.

Whay, H. R., Waterman, A. E., Webster, A. J. F. and O'Brien, J. K. (1998). The influence of lesion type on the duration of hyperalgesia associated with hindlimb lameness in dairy cattle. *Vet. J.* 156(1), 23–29.

Whay, H. R., Webster, A. J. F. and Waterman-Pearson, A. E. (2005). Role of ketoprofen in the modulation of hyperalgesia associated with lameness in dairy cattle. *Vet. Rec.* 157(23), 729–733.

Willis, G. L. (1983). A possible relationship between the flinch, step and kick response and milk yield in lactating cows. *Appl. Anim. Ethol.* 10(4), 287–290.

Winder, C. B., LeBlanc, S. J., Haley, D. B., Lissemore, K. D., Godkin, M. A. and Duffield, T. F. (2016). Practices for the disbudding and dehorning of dairy calves by veterinarians and dairy producers in Ontario, Canada. *J. Dairy Sci.* 99(12), 10161–10173.

Winder, C. B., Miltenburg, C. L., Sargeant, J. M., LeBlanc, S. J., Haley, D. B., Lissemore, K. D., Godkin, M. A. and Duffield, T. F. (2018). Effects of local anesthetic or systemic analgesia on pain associated with cautery disbudding in calves: a systematic review and meta-analysis. *J. Dairy Sci.* 101(6), 5411–5427.

Yousef, M. K. (1985). Thermoneutral zone. In: *Stress Physiology in Livestock Voume 1. Basic Principles*. CRC Press Press Inc, Boca Raton, FL, USA.

Part 2

Welfare indicators and monitoring

Chapter 4

Developing effective welfare measures for cattle

Ute Knierim, University of Kassel, Germany; Christoph Winckler, University of Natural Resources and Life Sciences Vienna, Austria; and Luc Mounier and Isabelle Veissier, Université Clermont Auvergne, INRAE, VetAgro Sup, France

1 Introduction

Welfare is a multi-dimensional concept. It includes any aspect of the animal's life that makes it a good life to live (see for instance the definition of five freedoms by the Farm Animal Welfare Council 1992). The assessment of the welfare of an animal or a group of animals can therefore hardly rely on one single measure (Botreau et al. 2007). Rather a series of measures should be used to check the various dimensions of animal welfare.

There has been extensive discussion on whether measure x or y is valid to assess animal welfare or whether the assessment of welfare shall be performed with resource-based measures (such as space allowance, quality of the management) versus animal-based measures (such as body condition, clinical symptoms, performance or behaviour). We argue that welfare measures should undergo a precise process of validation before being considered valid or not.

The objectives of this chapter are to draw a list of performance-characteristics that a welfare measure should possess in order to be considered valid for the assessment of animal welfare. Further, we aim to present a choice of validated measures that can be used to assess the welfare of cattle and to

http://dx.doi.org/10.19103/AS.2020.0084.05

discuss ways how they can be collected in practice. As a preamble, we will remind the reader of the various definitions of animal welfare, because they affect the measures chosen.

2 Definition of animal welfare

The Brambell committee established by the British government provided a first definition of animal welfare, by declaring that 'Welfare is a wide term that embraces both the physical and mental wellbeing of the animal' (Brambell 1965). In the debate on animal welfare concepts, this very general definition has been broken down into aspects that contribute to good or poor welfare. Diverging points of view then appeared on the relative importance of these aspects. There is apparently a consensus that a prerequisite for good welfare is the absence of intense and prolonged suffering from pain, hunger, thirst, discomfort, fear or stress. There is, however, no clear agreement on other less-severe issues around biological functioning (absence of malnutrition, injury and disease) or the experience of positive experiences (Fraser 1995). Positive experiences can come from access to particularly appreciated resources (for sleeping comfort, food) or from the expression of activities for which the animals are motivated (play activities such as calves frolicking with their tails up, exploration of the environment like pigs rooting in the ground, or cohesive social interactions like licking between animals that have affinity bonds) (Boissy et al. 2007). Fraser (1995) emphasises that increasing biological knowledge may move differing viewpoints closer to agreement, but that 'the inherently subjective element in judging the relative importance of different attributes' cannot be eliminated.

A core issue is that animals are sensitive beings, that is, they can experience affective states such as emotions, and for some authors that is why welfare makes sense (Duncan 2002, Veissier and Boissy 2007). According to these authors, it is not, for example, the disease that in itself degrades the welfare of an individual, but the fact that this individual feels sick. Here the emphasis is on mental health, and physical health is supposed to affect mental health. Discussing whether welfare is solely a matter of mental state or a matter of both mental and physical health is out of scope of the present chapter. Actually, the assessment of animal welfare is generally based on assessing the behaviour of an animal - that may reflect its mental state - and its physical state - either because it is considered to be part of welfare or because it may affect the mental state.

The different aspects of animal welfare, or principles that must be fulfilled to assure a good life, can serve as a basis to develop checklists for the assessment of animal welfare. The Farm Animal Welfare Council, which followed the Brambell Committee, defined five principles to be respected in order to achieve good welfare; their descriptions put emphasis on what animals may feel (Farm Animal Welfare Council 1992):

- Absence of prolonged hunger and thirst;
- Physical comfort;
- No injuries, illnesses or pain;
- Absence of fear and distress; and
- Ability to express the normal behaviour of the species.

These principles, commonly known as the *five freedoms*, form the basis of many regulations and methods of assessing welfare. They were detailed into 12 independent criteria in the Welfare Quality® project in order to propose a common grid, making it possible to identify a set of measures covering all aspects of welfare (Botreau et al. 2007):

- Absence of prolonged hunger: Animals should not suffer from prolonged hunger, that is, they should have a suitable and appropriate diet;
- Absence of prolonged thirst: Animals should not suffer from prolonged thirst, that is, they should have a sufficient and accessible water supply;
- Comfort around resting: Animals should have comfort when they are resting;
- Thermal comfort: Animals should be neither too hot nor too cold;
- Ease of movement: Animals should have enough space to be able to move around freely;
- Absence of injuries: Animals should be free of injuries, for example, skin damage and locomotion disorders;
- Absence of disease: Animals should be free from disease;
- Absence of pain induced by management procedures: Animals should not suffer pain induced by inappropriate management, handling, slaughter or surgical procedures (e.g. castration, dehorning);
- Expression of social behaviours: Animals should be able to express normal, non-harmful, social behaviours (e.g. grooming);
- Expression of other behaviours: Animals should be able to express other normal behaviours, that is, it should be possible to express species-specific natural behaviours such as foraging;
- Good human-animal relationship: Animals should not be afraid of humans and be handled well in all situations, that is, handlers should promote good human-animal relationships; and
- Positive emotional state: Negative emotions such as fear, distress, frustration or apathy should be avoided whereas positive emotions such as security or contentment should be promoted.

With such operational definitions of animal welfare, measures can be proposed and problems quantified (Stafleu et al. 1996). Indeed, the Welfare Quality® framework is now being used as a reference to develop animal welfare assessment systems in various species, that is, cattle, pigs, hens and broilers during the

Welfare Quality project but also horses, sheep, turkeys, and dolphins (Welfare Quality 2009, AWIN 2015, Clegg et al. 2015). In these assessment systems, the measures to check the various welfare aspects were chosen according to their validity, their ability to measure what is aimed to be measured with an acceptable precision, and their feasibility, that is they can be applied in field conditions.

3 Performance characteristics to define valid welfare measures

In animal welfare science, the validation process concerning measures is commonly broken down into the assessments of 'validity' and 'reliability', with different possible approaches such as determining face, concurrent or construct validity (see below) and testing intra-observer or inter-observer reliability (e.g. Scott et al. 2001, Keeling 2009). To go further and formalise the validation process, we looked towards further disciplines and their approaches of validation to see what can be learned for the animal welfare case. For instance, for analytical assays (e.g. biochemical or microbiological assays), the validation of a method (test of fitness for purpose) includes the evaluation of method performance-characteristics such as selectivity, trueness, precision, working range, sensitivity, ruggedness and matrix variation (Taverniers et al. 2004, Magnusson and Örnemark 2014). In addition, the measures need to be fit for use, that is practical aspects such as acceptability by scientific and regulatory communities (which might conform to face validity) and feasibility, including cost and time needed, must also be taken into account (EFSA AHAW Panel (EFSA Panel on Animal Health and Welfare) 2012).

Not all performance criteria of analytical methods might be applicable to animal welfare. Here, we propose a validation process inspired from these concepts so as to strengthen the choice of welfare measures in further studies. Although terms can vary between disciplinary fields, we preferably use the definitions of performance characteristics provided by Magnusson and Örnemark (2014) in the Eurachem Guide.

3.1 Selectivity and trueness

For analytical methods, selectivity is 'the extent to which the method can be used to determine particular analytes in mixtures or matrices without interferences from other components of similar behaviour' (Vessman et al. 2001). For a welfare measure, selectivity may first refer to the degree to which a measure can quantify what we want to analyse and not something else. This property should be analysed in reference to the welfare criterion the measure is supposed to bring information about, for example, 'is a method valid to detect stress, abnormal behaviour or hunger, etc.'? It is often difficult to distinguish selectivity from trueness, which – for analytical methods – refers to the closeness of agreement between a test result and the accepted reference value of what is being measured (Thompson et al. 2002).

These two properties are often considered together and for welfare measures (lacking a reference value) they can be assessed by (Scott et al. 2001, Reenen and Engel 2004):

- Comparing the results produced by the method and those produced by another already validated method that serves as a gold standard (concurrent or criterion validity). For instance, regarding behaviour, one may compare the results obtained during a short period to those obtained by longer observations. In case scan sampling is used (i.e. one snapshot observation every fixed interval of time), the results can be compared to those obtained with continuous observations.
- Comparing the effects of conditions or treatments or demonstrating causal relationship between treatment and effect (predictive or construct validity). For instance, regarding the criterion 'Good human-animal relationship', one may ask whether an approach test - measuring at what distance one can approach an animal before it moves away - performed when the animal is feeding reflects the quality of the human-animal relationship. This question was answered in calves by comparing animals that received positive versus negative contacts with humans: the former accepted to be approached at a closer distance than the latter (Lensink et al. 2003).
- Agreeing among experts on the validity of a measure according to their experience (consensus or face validity), if concurrent or construct validity cannot be established.

It is also necessary to check for confounding factors that can influence the results. For instance, lameness can affect the results of an approach test: lame cows can be approached by a shorter distance than healthy ones, probably because the motivation to avoid humans is counterbalanced by the pain caused by the movement (Špinka et al. 2005). In that case, lameness can be a confounding factor.

3.2 Precision

For analytical methods, precision refers to how close results are to one another (Magnusson and Örnemark 2014). It is often expressed as standard deviation or relative standard deviation. It includes:

- Repeatability (sometimes called intra-assay repeatability), that is, the closeness between measures done in the same conditions, that is, by the same operator and with the same equipment.

- Reproducibility (sometime called inter-assay repeatability), that is, the closeness between measures done in different conditions, for example, by different operators, or with different equipment, at different times.

For welfare measures, authors often refer to:

- *Intra-observer repeatability (or intra-observer reliability)*, that is, the similarity between measures done by the same observer under similar conditions. This can be tested using video recordings for behaviours or assessing the same animals again after a short time lapse for clinical symptoms.
- *Inter-observer repeatability (or inter-observer reliability)*, that is, the similarity between measures done by several observers, under the same conditions. This may be checked by asking several observers to look at the same animals at the same time.

Test-retest repeatability (or test-retest reliability) is also sometimes used, meaning that a method should produce same results when it is applied twice within a short period of time. This property rather refers to ruggedness, which is described below.

Very often, repeatability is expressed as a correlation coefficient between observations in case of numerical data, for example, frequencies of certain behaviours. Spearman, Pearson or intra-class correlations, or Kendall's coefficient of concordance for more than two observers are then calculated. In case of ordinal or nominal data, for example, lesion scores, often kappa coefficients are used (e.g. PABAK: prevalence-adjusted and bias-adjusted kappa). Although there is no natural limit when repeatability is high enough to produce trustworthy data, a correlation of above 0.70 or a concordance of more than 0.40 is generally considered as a minimum when the measure is very important and it is difficult to reach higher agreements. However, values above 0.8 for a correlation or 0.6 for a concordance should be the goal (Fleiss et al. 2003, Martin and Bateson 2007, Knierim and Winckler 2009).

3.3 Working range and sensitivity

Sensitivity is a term used with a number of differing meanings (Magnusson and Örnemark 2014). While diagnostic sensitivity refers to the ability to correctly identify true positive cases (e.g. a disease) from all assessed cases, analytical sensitivity describes how well a change in analytical results corresponds to change in the measured quantity (e.g. an analyte concentration) (Feinberg 1996). Sometimes the term is also used in the sense of limit of detection. Thus, it is always important to qualify the intended meaning of the term. Regarding analytical sensitivity, one may question which degree of differentiation of a welfare measure is necessary. Is it, for instance, sufficient to differentiate between minor versus severe lameness or mild versus severe wounds? Insufficient differentiation may level farms with truly different welfare states, while very high differentiation may negatively affect precision (i.e. repeatability, reproducibility, or reliability) of the measurement.

For analytical methods, the working range is 'the interval over which the method provides results with an acceptable uncertainty' (Magnusson and Örnemark 2014). This interval should match with the purpose of the end-user of the method. In particular, the limit of quantification, that is the smallest value that can be detected correctly, defines the lowest end of the working range. For our purpose, the limit of quantification should allow detecting a welfare problem, for example, an animal does not suffer from hunger or is not diseased. To assess whether the quantification limit of a measure is fit for the purpose, we may question whether the method can be used to detect slight welfare problems. For instance, the indicator 'sunken eyes' is used to detect dehydration in calves. This however indicates an extreme state where this animal urgently needs to be rehydrated, often by intravenous administration of an electrolyte solution. This symptom is not sensitive enough to detect thirst that may occur on a farm, if the water points are scarce and the animal needs to wait a long time before getting access to water. Nevertheless, the symptom of sunken eyes appears appropriate to detect dehydration after very long transports during warm weather. Another example regarding the quantification limit relates to the occurrence of rare behaviours (such as certain abnormal behaviours or play in adults). This could be detected by observing animals, but not necessarily be reliably quantified within a limited time frame (Knierim and Winckler 2009).

3.4 Ruggedness

In terms of analytical method validation, the ruggedness (sometimes called robustness) is the capacity of a method to produce similar results when minor deviations are made from the experimental procedure (Magnusson and Örnemark 2014). This can be assessed by making deliberate changes to the procedure and assess the impact on the results.

For welfare measures, this may be transposed to *stability over time*, especially day-to-day variations. Indeed, the results of a welfare assessment should be representative of the long-term farm or slaughter-plant situation. Although the overall conditions on a farm may not change during a given period and the overall welfare of animals should thus be stable, there might be differences in the results obtained from some measures because the conditions for observation have slightly changed or because of random variations between days. This is particularly true for behaviours whose occurrence is determined by multiple factors. Weather conditions or occasional events like small changes in the routine management of the farm can affect the expression of behaviours. The impact of such changes on the results should be analysed. The method will be considered robust if similar recordings are achieved at different times if no major changes on the farms - that is, changes that are significant for the welfare of the animals - occurred (Knierim and Winckler 2009).

3.5 Matrix variation

For analytical methods, the matrix is the set of constituents present in the material on which the method is applied. These constituents may have an effect on the results of the method (Feinberg 1996, Thompson et al. 2002). For instance, the determination of hormones in different tissues or species may require specific methods.

For welfare measures to be applied on farms or at slaughter, it is important to check if the measure can be applied in different farming or slaughter systems. For instance, detecting lameness in cows in a loose barn versus tie-stalls requires a different method. In Welfare Quality® lameness is assessed in loose-housed cows by making them walk in a straight line and checking if the cow bears its weight equally on the four limbs and makes regular steps. For tied cows, the observer checks if the cow stands on its four limbs when undisturbed. Then the observer makes the cow move to the left and to the right, observing how she shifts weight from foot to foot (Welfare Quality® 2009).

3.6 Fitness for use

When the assessment has to be performed on a large scale and under commercial conditions, the feasibility of a method is essential. Important issues are the time necessary to carry out a measure (e.g. long-term observations to detect changes in time budget are less feasible), the need for specific devices to perform the measure, the requirement for specific skills to perform the measure (e.g. for taking blood samples) or the cost of measures. Feasibility relates to the knowledge and devices available at present. Therefore, a measure that is considered not feasible today may become feasible in the future, for instance, thanks to automatic recordings becoming widely applicable (e.g. automatic recordings of animals' movements).

We believe that following such a validation process for welfare measures is likely to help make such measures more acceptable by a wide range of users, thus helping to improve animal welfare.

4 Choice of measures for cattle

As described previously, important aspects to consider when choosing measures of welfare are validity, reliability and feasibility. In this section, we first explain the selection process using two examples from the Welfare Quality® assessment protocol for dairy cattle. Then we discuss the main drivers for the composition of assessment protocols.

The measures included in the Welfare Quality® protocol for dairy cattle are listed in Table 1 together with a semi-quantitative assessment of the respective levels of validity, reliability and feasibility. For the criterion 'Expression of social behaviours', both agonistic and affiliative behaviours were considered for the sake of content validity, that is, in order to include all relevant items of social behaviours as a welfare measure. For feasibility reasons, behaviour observations had to be limited to 2 h to allow completion of the on-farm assessment within a day. Within this time frame, on-farm observations in different housing systems and countries revealed good to very good inter-observer agreement and at least an acceptable stability over time for the incidence of behaviours such as head butts or displacements, as well as for the incidence of total agonistic interactions (additionally including fighting, chasing or chasing up from the lying area) (Laister et al. 2009a,b). Affiliative behaviours such as social licking were considered as potential measures of good welfare. However, the validity of licking as a herd measure of positive social interactions is questionable because it may just alleviate poor welfare due to social tension (Knierim and Winckler 2009). It was eventually excluded as a measure of socio-positive behaviour from the operational assessment protocol.

Table 1 shows that valid animal-based welfare measures were not available for all criteria. For example, for absence of thirst, the skin test, which uses the delay needed for the skin to resume its initial position after being pinched, has been shown lacking relation to serum osmolarity or packed cell volume (indicating the level of hydration) and to drinking behaviour (Pritchard et al. 2006). For the direct measurement of osmolarity and haematocrit counts (Knowles et al. 1995, Pritchard et al. 2006), more research is needed to establish if these indicators can help to distinguish a moderate dehydration from a severe one. Additionally, collection of a blood sample on farms is an invasive procedure and time-consuming. In view of the above concerns, recording of resource-based measures such as the number of water points, the flow of water

Table 1 Welfare principles, criteria, individual measures and type of validity available for the measures in the Welfare Quality® assessment protocol for dairy cattle

Principle	Criterion	Welfare measure	Type of validity[1]	Inter-observers repeatability[2]	Time needed to perform a measure
Good feeding	Absence of prolonged hunger	Body condition	Concurrent validity (culling risk, infertility)	++	3 min/animal (incl. assessment of cleanliness, integument alterations, lameness, nasal/ocular/vulvar discharge, signs of diarrhoea)
	Absence of prolonged thirst	Resource-based (provision of water)	Face validity	Good (cleanliness of trough) to very good	15 min/unit
Good housing	Comfort around resting	Time needed to lie down	Face validity (e.g. unpleasant experience of hard surface)	+++	150 min (incl. observations of agonistic behaviour)
		Animals lying partly/ completely outside	Face validity (unpleasant experience, injuries, soiling)	+++	
		Cleanliness	Face validity (skin inflammation, itching)	+++	See absence of prolonged hunger
	Thermal comfort	No measure available yet	-	-	-
	Ease of movement	Resource-based (housing system)	Face validity	Very good	~15 min (interview, incl. somatic cell count, management procedures, pasture access)
Good health	Absence of injury	Lameness	Construct validity (analgesics lead to lower lameness score)	++	See absence of prolonged hunger
		Integument alterations	Face validity (pain)	++	
	Absence of disease	Somatic cell count	Concurrent validity (inflammation, pain associated with mastitis)	+++	See ease of movement

		Coughing, nasal discharge	Face validity (respiratory disorder)	++	See absence of prolonged hunger
		Vulvar discharge	Face validity (uterine inflammation)	++	See absence of prolonged hunger
	Absence of pain induced by management procedures	Management-based (methods used for disbudding/ dehorning, castration, tail docking)	Concurrent validity (behavioural and physiological indicators of pain)	Very good	See ease of movement
Appropriate behaviour	Expression of social behaviours	Agonistic behaviours such as head butts, displacements	Concurrent validity (unpleasant, stressful situations, risk of injuries)	+++	150 min (including observations of behaviour around resting)
	Expression of other behaviours	Resource-based (access to pasture)	Concurrent validity (positive effects on e.g. lameness, skin alterations, cleanliness, mortality)	Very good	See ease of movement
	Good human-animal relationship	Avoidance distance towards human	Construct validity (quality of handling)	+++	20 min
	Positive emotional state	Qualitative behaviour assessment	Face, concurrent and construct validity (overall impression, association with quantitative measures of behaviour)	+/++	25 min

1: concurrent/construct validity for experimental setups and specific test situations only, not shown for on-farm assessment of groups of animals.
2: +++: Kappa, Kendall's W, $r>0.8$; ++: Kappa = 0.6–0.8, Kendall's W, $r=0.7$–0.8; +: Kappa = 0.4–0.6.

and its cleanliness are currently considered more appropriate for assessing compliance with the criterion 'Absence of prolonged thirst'.

Although the Welfare Quality® protocols aim at allowing for a comprehensive assessment covering all dimensions of welfare, a number of main behavioural measures, for which at least face validity can be established, were not included. For example, this applies to play behaviour (Fregonesi and Leaver 2001), abnormal behaviour such as tongue rolling or intersucking (Krohn 1994, Lidfors and Isberg 2003) and comfort behaviours such as brush use (Mandel et al. 2016). All these measures require long-term direct observations for a reliable assessment and are therefore often considered not feasible in the context of on-farm assessments (Knierim and Winckler 2009).

Comparing welfare assessment protocols which are implemented in the dairy industry (Table 2) reveals marked differences in the number and type of measures included. Measures of health and physical appearance - mainly addressing the biological functioning view of welfare - are consistently included. This is less often the case for behavioural measures that provide information about affective states or the ability to perform normal behaviour. Protocols also differ substantially from each other in their comprehensiveness regarding health measures. Lameness, body condition and skin alterations appear to be 'core measures'. More detailed information on the health state is less-frequently obtained.

Feasibility in terms of time needed and associated costs is often mentioned as the main driver for the final decision on which measures to include in assessment protocols (Sorensen et al. 2007, Metz et al. 2015). Some limitations have to be accepted for on-farm assessments, which - as mentioned above - may lead to the omission of welfare relevant indicators such as play behaviour. The ultimate aim of the assessment also determines the content of protocols. For example, the Welfare Quality® protocols aim at providing a comprehensive picture of welfare by addressing the four principles 'Good feeding', 'Good housing', 'Good health' and 'Appropriate behaviour' (see Table 1). They have initially been developed to increase transparency of production, to integrate animal welfare in the food quality chain and to promote welfare improvement (Blokhuis et al. 2010). Farm assurance schemes may, however, only target health-centred core problems of the industry as in the case of the US-based FARM programme, or take a slightly broader approach like the AssureWel protocols which are now used by the UK-based Soil Association (organic farming) and RSPCA Assured (welfare label). The association of health-related welfare issues with productivity and thus economic benefits may further explain the focus on health-centred measures. Respective outcomes of the assessment can also be used for decision support in herd management (Winckler 2019). However, it remains unknown whether such health-centred approaches sufficiently take societal expectations into account, which typically relate to naturalness for dairy cattle (Beaver et al. 2020).

5 Collection of animal-based welfare measures on the farm

As initially explained, the number of welfare measures is large due to the multifactorial nature of welfare, that is, because many welfare aspects need to be covered. In addition, almost all welfare aspects can be assessed in different ways regarding the kind of measure and the method of recording. For instance, the prevalence and the severity of lameness in a herd can be ascertained by visual scoring of the walking gait of a sample of animals that are induced to walk a certain route (e.g. Welfare Quality® 2009) or by automatic means (reviewed by Alsaaod et al. 2019). Alternatively, claw trimming or treatments of lame cows can be documented in farm records or the assessments of claw condition can be carried out at the slaughterhouse, for example, in beef bulls (e.g. Magrin et al. 2019). In this section, the pros and cons of different methods of data acquisition are discussed with reference to performance criteria explained above.

5.1 Welfare measures obtained from direct observations or assessments on the farm

Currently, the direct observation or assessment of the animals on-farm is an important component of most welfare assessments (e.g. Welfare Quality® 2009). To obtain precise (i.e. reliable) data, the assessors must be well trained and inter-observer repeatability needs to be regularly tested (Knierim and Winckler 2009). However, repeatability testing is not profane. It needs a sufficient number of samples that are as independent as possible (as a rule of thumb it should be at least 10, but preferably more). For practical reasons, some compromises are usually made. For instance, when the outcomes are prevalences of certain health conditions on a farm, one should ideally test inter-observer repeatability regarding 10 farm prevalences or more. In practice, one will rather check repeatability at animal level rather than farm level. Usually in this case, the number of observational units can be much higher than 10 animals which is advantageous in terms of power of the reliability testing. Further examples of compromises relate to behavioural observations of groups of animals, where several places in one house or different observation times may be used as observational units, even though they are not truly independent from each other. Finally, pictures (e.g. of integument alterations of various severity), or videos (e.g. of social interactions) can be used to supplement reliability testing. Although some differences compared to live assessments can be expected,

Table 2 Measures included in protocols for cows currently being used in the dairy industry: Welfare Quality® protocol for cattle (http://www.welfarequalitynetwork.net/network/45848/7/0/40); FARM, National Dairy Farmers Assuring Responsible Management Program (https://nationaldairyfarm.com/farm-animal-care-version-4-0/); AssureWel, Advancing Animal Welfare Assurance (http://www.assurewel.org/dairycows) (modified after Winckler 2018)

	Measure	Welfare Quality	FARM	AssureWel
Physical appearance and health	Body condition	x	x	x
	Cleanliness	x		x
	Skin alterations	x[1]	x[2]	x[1]
	Broken tails		x	x
	Lameness	x	x	x
	Mastitis	x		x
	Respiratory signs	x		
	Diarrhoea/loose faeces	x		
	Vulvar discharge	x		
	Downer cows	x		
	Animals needing further care			x
	Mortality: unplanned culls/casualties	x		x
Behaviour	Agonistic behaviours	x		
	Behaviour around resting	x[3]		
	Human-animal relationship	x[4]		x[5]
	Qualitative behaviour assessment	x		

[1] hair loss, lesions, swellings at one side of the animal.
[2] hock/carpal joint injury.
[3] time needed to lie down, lying partly/completely outside the lying area.
[4] avoidance distance towards unknown person.
[5] response to stockperson.

such a procedure has the advantage that a more complete range of conditions can be assessed than on a farm. Indeed, all severity scores of health conditions and all behavioural measures provided for in the assessment protocol should be covered by the testing. For statistical reasons, different scores should be represented as evenly as possible, and zero values in metric data (for instance no agonistic behaviour) should not be over-represented. To achieve this, considerable organisational efforts (often including the use of videos or pictures) and sufficient time are commonly required and must be planned for. Also in scientific work, proper repeatability testing needs to be extended (e.g.

for behavioural observations: Kaufman and Rosenthal 2009, Burghardt et al. 2012).

Direct on-farm assessments are often limited by time and labour and thus financial constraints. This can affect how representative the recorded data are of the farm situation. Sample size is one of the difficult issues. In most cases, health conditions cannot be assessed in all animals, but only in a representative sample. For instance, Welfare Quality® (2009) recommends sampling for 73 animals in a herd of 300. Many claim that such a number usually overstrains available resources and disrupts farm routine unduly. However, others regard it insufficient for the reliable estimation of a prevalence. According to the formula provided by Cochran (1977), with 73 out 300 animals, the estimation may deviate ±10% from the true prevalence (with 95% confidence level) when supposing a prevalence around 50% (if the true prevalence is higher or lower, precision will become higher, thus 50% represents the 'worst case'). This 10% precision should be taken into account when comparing and interpreting results. For behavioural observations, the duration of observation should be long enough to fairly represent activities over a longer period. For instance, Laister et al. (2009a) compared frequencies of agonistic interactions during 4 h and 2 h of observation time after feeding and concluded that due to high correlations ($r>0.8$) the shorter observation time would be sufficient. However, within short observation times it will not be possible to reliably record rare behaviours that occur unpredictably and unevenly over time (see for instance Plesch et al. 2010 for resting behaviour).

Another difficult issue refers to the stability of snapshot assessments or observations over time, that is, whether they are representative of the general farm situation. Seasonal influences, for instance, from grazing, are to be expected. Moreover, short-term events such as routine claw-trimming of the herd, herd treatments, or recent regrouping or introduction of new animals to the herd can profoundly affect the outcome of a welfare assessment. Behavioural measures are especially sensitive to such changes and they can even be biased by the presence of an unfamiliar observer. By contrast, health measures are expected to reflect medium- to long-term influences and therefore to show less short-term fluctuations. However, when Winckler et al. (2007) assessed, among others, lameness, tarsal joint lesions and soiling of hindquarters in cows every two months on eight dairy farms, they found varying correlations between prevalences from 0.48 to 0.83. Only correlations between results from single assessments and the average of all assessments showed coefficients of at least 0.70 with only few exceptions. Also Kirchner et al. (2014), applying the Welfare Quality® assessment protocol on 63 beef bull farms repeatedly after one and six months, found only few consistencies regarding animal-based measures after one month and none after six months. As a general conclusion therefore, usually repeated measurements over time are necessary for a reliable assessment of the long-term welfare level on a farm.

A final reliability aspect is the avoidance of an expectation bias (Tuyttens et al. 2014). In direct observations blinding of the assessor is not possible. Automatic or slaughterhouse recordings have an advantage in this regard. In addition, their implementation may be less labour-intensive. However, these methods also have their specific disadvantages as outlined below.

5.2 Welfare measures obtained from farm documentation or production records

Some health conditions such as certain infections require either diagnostic procedures - which are commonly not available in the context of an on-farm assessment - or longer observation periods because they occur infrequently. Farm records of treatment incidences are often proposed as measures (e.g. Whay et al. 2003). However, such data need to be used carefully. The decision when to treat a diseased animal may greatly vary between farms. A lack of treatment does not necessarily indicate the absence of welfare problems. Instead, for example, refraining from antibiotic treatment may be due to imminent slaughter, a culling strategy, or because of label restrictions. In addition, the quality of the farm records in terms of completeness, reliable diagnosis, and distinction between prophylactic and therapeutic treatments is often insufficient. Similar considerations apply to mortality or culling rates. The decision when to cull an animal relates only to some degree to the health or welfare state of the animal. Market considerations (current prices for milk or meat, changes in subsidy programmes), breeding or eradication programmes may greatly affect this decision. Some countries, namely Scandinavian countries and Austria, run centrally organised health databases which helps to increase reliability of documentation, but possible influences of factors not related to welfare still need to be considered when interpreting these data.

A useful source of information are measures from milk performance testing, such as somatic cell count in milk (as indicator of mastitis) or fat-to-protein ratios (indicating risks of metabolic disorder). Again, care is needed in the choice of specific measures and their interpretation in terms of animal welfare. For example, temporary increases in cell counts may not always be disease-related.

Farm-routine data are owned by the farmer. Depending on the framework of the welfare assessment this may pose limitations. Such data are a priority source of information to the farmers for self-assessment of animal welfare to improve the management of the farm. However, they may not be available for third-party auditing. Their external use needs the farmer's permission, and a very careful data handling is necessary in order to protect privacy of the owner. This similarly applies to automatically recorded data on farm and slaughter records.

5.3 Welfare measures automatically recorded on the farm

Technical progress increasingly enables the use of sensors and evaluation algorithms (e.g. using artificial intelligence) to monitor a great number of health conditions or behaviours. For instance, over the last two decades, a multitude of automated methods have been developed for the detection of lameness (reviewed by Alsaaod et al. 2019). They include image-processing techniques, use of accelerometers, pressure-sensitive walkways, ground reaction force systems, four-scale weighing platforms, auditory signal analysis and indirect methods such as thermography and use of behavioural measures like different measures of feeding behaviour, frequency of visits to automated milking systems or milk production, sometimes using a combination of different measures. Alsaaod et al. (2019) list 49 peer-reviewed validation studies on this topic. It is claimed that the automated recording of lameness may be more sensitive and less prone to observational bias compared to observation by a human (Rushen et al. 2012, Alsaaod et al. 2019). Nevertheless, its application in practice lags behind. This is likely due to a number of factors: it is a long and expensive process to reach a fully automated system that produces not only raw data, but also reliable evaluations (e.g. lame/not lame). Only part of the studies listed by Alsaaod et al. (2019) has reached this final stage. The resulting high initial price of such techniques reduces the readiness of farmers to adopt them. Moreover, to date the few commercially available solutions are not completely satisfactory with regard to, for example, sensitivity (Bicalho et al. 2007). Some of the techniques developed, for example, the use of accelerometers, are only partly suitable for everyday use on farms. The equipment mounted on the animal may affect their behaviour, and even impair their welfare by being uncomfortable or causing lesions. Limited battery capacities are another issue to be solved. Moreover, the devices may get lost, which is especially problematic when they are expensive. If they shall be used for third-party assessments, extra efforts for mounting and removing the devices may render a direct assessment more time efficient. When data from farm-owned automatic monitoring systems are to be used for third-party auditing or benchmarking between farms, then the comparability of results from different systems needs to be ascertained. This may be difficult, because sensitivity and specificity vary between systems and are affected by individual farm factors. Among them are illumination levels (for image processing), cleanliness of sensors, background noise (for acoustic processing) or individual technical configuration. Truly, repeatability testing on each farm would be necessary. Often this issue is underestimated owing to the belief that automated measurements are 'objective', that is, not biased by human perception. This is often true, but potential bias by further external factors should not be overlooked.

A further advantage of automated systems is the increased possibility to include behavioural measures into on-farm welfare assessments, both on individual and group level (Rushen et al. 2012). This may relate to, for example, lying, feeding, play and other behaviours. In addition, long-term recording of activities indicative of disease, such as coughing in relation to pneumonia in calves, will provide more reliable results than short-term direct recording of coughing. The limited time available during direct observations is one reason why currently most animal welfare protocols show a clear dominance of health measures which could be overcome at least partly by adoption of automatic recording systems.

5.4 Welfare measures obtained from slaughter records

Disease diagnoses from the official or company control in slaughterhouses can be used as welfare measures for categories of cattle slaughtered in groups, that is, the prevalence of pneumonia in veal calves. In theory they are especially valuable, because each slaughtered animal is assessed by an expert at the slaughterhouse. However, currently there is insufficient training and reliability testing of the assessors, so that outcomes ascertained by different assessors and, even more so, in different slaughterhouses are hard to compare. Another disadvantage of welfare measures from slaughter is their retrospective nature; they cannot be used to remedy welfare problems in the animals affected, but only for the next animals that will live on the farm. We nevertheless suggest that automatic recording of welfare measures at the slaughter line has great potential which is nowadays used in poultry but should be further developed in cattle.

6 Conclusion

There is no unique measure for animal welfare but rather sets of measures, especially related to health and behaviour, which are to be used to address the various aspects of welfare.

As for analytical methods, it is possible to validate these measures by characterising their performances in terms of selectivity, trueness, precision, and so on. Such a process was followed to design the Welfare Quality® protocols, which include measures for which the validity, reliability (precision and ruggedness) and feasibility were deemed adequate. We acknowledge that in some rare cases; there is at present no measure sufficiently validated so that the least objectionable measures available at present have to be used (e.g. access to water instead of addressing thirst) or a specific aspect is not addressed at present (e.g. thermal comfort).

Each way of recording welfare measures in practice has pros and cons; knowing them as well as the specific challenges of different methods allows an informed choice and helps to avoid possible pitfalls. Because advantages or disadvantages of methods depend on the specific measure to be recorded, often a combination of different methods is a good solution. Lastly, the development of sensors on farms opens possibilities to monitor continuously the health and behaviour of animals and may make animal welfare measurement more viable in the future.

7 Where to look for further information

At present, not all animal welfare criteria can be checked easily in field conditions. There is, for example, a need to develop and validate measures to check the absence of thirst and the thermal comfort of cattle. In addition, most welfare assessment methods are focused on animals indoors. Cattle grazing outdoors may nevertheless be exposed to welfare hazards and their welfare should be checked adequately. Sensors offer opportunities to observe animals 24/7; however, more work is needed to be able to comprehensively assess animal welfare using sensors.

There are numerous research teams working on animal welfare over the world. The Welfare Quality Network gathers partners from the former Welfare Quality project. It aims to maintain the Welfare Quality® protocols updated according to the most recent technical and scientific knowledge. The conference Assessment of Animal Welfare at Farm and Group Level (WAFL) is held every 3 years with the aim to strengthen connections between animal welfare scientists around the world. Both the Welfare Quality Network and the WAFL conference can be a source of inspiration for those planning to design a welfare assessment tool.

8 Acknowledgements

Part of this work was supported by the Swedish Centre for Animal Welfare (Project 'ENCAW Pilot Study: Impact Assessment'). The work from Isabelle Veissier, coordinator of the present chapter, is supported by the Agence Nationale de la Recherche of the French government through the programme "Investissements d'Avenir' (16-IDEX-0001 CAP 20-25 2017).

9 References

Alsaaod, M., Fadul, M. and Steiner, A. 2019. Automatic lameness detection in cattle. *Vet. J.* 246, 35–44.

AWIN. 2015. *AWIN Welfare Assessment Protocole for Horses*. 80 pp. DOI: 10.13130/AWIN_HORSES_2015.

Beaver, A., Proudfoot, K. L. and von Keyserlingk, M. A. G. 2020. Symposium review: Considerations for the future of dairy cattle housing: An animal welfare perspective. *J. Dairy Sci.* 103(6), 5746–5758.

Bicalho, R. C., Cheong, S. H., Cramer, G. and Guard, C. L. 2007. Association between a visual and an automated locomotion score in lactating Holstein cows. *J. Dairy Sci.* 90(7), 3294–3300.

Blokhuis, H., Veissier, I., Miele, M. and Jones, B. 2010. Welfare Quality® and beyond. *Acta Agric. Scand. Sect. A Anim. Sci.* 60, 129–140.

Boissy, A., Manteuffel, G., Jensen, M. B., Moe, R. O., Spruijt, B., Keeling, L. J., Winckler, C., Forkman, B., Dimitrov, I., Langbein, J., Bakken, M., Veissier, I. and Aubert, A. 2007. Assessment of positive emotions in animals to improve their welfare. *Physiol. Behav.* 92(3), 375–397.

Botreau, R., Veissier, I., Butterworth, A., Bracke, M. B. M. and Keeling, L. J. 2007. Definition of criteria for overall assessment of animal welfare. *Anim. Welf.* 16, 225–228.

Brambell, F. W. R. 1965. Report of the Technical Committee to enquire into the welfare of animals kept under intensive livestock husbandry systems. Command Paper 2836, Her Majesty's Stationery Office, London.

Burghardt, G. M., Bartmess-LeVasseur, J. N., Browning, S. A., Morrison, K. E. and Stec, C. L., Zachau, C. E. and Freeberg, T. M. 2012. Perspectives – minimizing observer bias in behavioral studies: A review and recommendations. *Ethology* 118, 511–517.

Clegg, I., Borger-Turner, J. and Eskelinen, H. 2015. C-well: The development of a welfare assessment index for captive bottlenose dolphins (Tursiops truncatus). *Anim. Welf.* 24(3), 267–282.

Cochran, W. G. 1977. *Sampling Techniques* (3rd edn.), New York: Wiley & Sons.

Duncan, I. J. H. 2002. Poultry welfare: science or subjectivity? *Br. Poult. Sci.* 43, 643–652.

EFSA AHAW Panel (EFSA Panel on Animal Health and Welfare). 2012. Scientific opinion on the use of animal-based measures to assess welfare of dairy cows. *EFSA J.* 10, 2554 [2581 pp.]. Available at: www.efsa.europa.eu/efsajournal.

Farm Animal Welfare Council. 1992. FAWC updates the five freedoms. *Vet. Rec.* 17, 357.

Feinberg, M. 1996. *La validation des méthodes d'analyse. Une approche chimiométrique de l'assurance qualité au laboratoire*, Paris: Masson, 108–124.

Fleiss, J. L., Levin, B. and Paik, M. C. 2003. *Statistical Methods for Rates and Proportions*, Hoboken, NJ: John Wiles & Sons.

Fraser, D. 1995. Science, values and animal welfare: Exploring the 'inextricable connection'. *Anim. Welf.* 4, 103–117.

Fregonesi, J. A. and Leaver, J. D. 2001. Behaviour, performance and health indicators of welfare for dairy cows housed in strawyard or cubicle systems. *Livest. Prod. Sci.* 68(2–3), 205–216.

Kaufman, A. B. and Rosenthal, R. 2009. Can you believe my eyes? The importance of interobserver reliability statistics in observations of animal behaviour. *Anim. Behav.* 78(6), 1487–1491.

Keeling, L. 2009. Defining a framework for developing assessment systems. In: *An Overview of the Development of the Welfare Quality® Project Assessment Systems. In Welfare Quality Reports 12*, pp. 1–7, Cardiff, UK: Cardiff University.

Kirchner, M. K., Westerath, H. S., Knierim, U., Tessitore, E., Cozzi, G. and Winckler, C. 2014. On-farm animal welfare assessment in beef bulls: Consistency over time of single measures and aggregated Welfare Quality((R)) scores. *Animal* 8, 461–469.

Knierim, U. and Winckler, C. 2009. On-farm welfare assessment in cattle: Validity, reliability and feasibility issues and future perspectives with special regard to the Welfare Quality® approach. *Anim. Welf.* 18, 451–458.

Knowles, T. G., Warriss, P. D., Brown, S. N., Edwards, J. E. and Mitchell, M. A. 1995. Response of broilers to deprivation of food and water for 24 hours. *Br. Vet. J.* 151(2), 197–202.

Krohn, C. C. 1994. Behavior of dairy-cows kept in extensive (loose housing pasture) or intensive (tie stall) environments. III. Grooming, exploration and abnormal-behavior. *Appl. Anim. Behav. Sci.* 42(2), 73–86.

Laister, S., Brörkens, N., Lolli, S., Zucca, D., Knierim, U., Minero, M., Canali, E. and Winckler, C. 2009a. Reliability of measures of agonistic behaviour in dairy and beef cattle. In Forkman, B. and Keeling, L. (Eds), *Assessment of Animal Welfare Measures for Dairy Cattle, Beef Bulls and Veal Calves*, pp. 95–112, Cardiff, UK: Cardiff University.

Laister, S., Brörkens, N., Minero, M., Lolli, S., Zucca, D., Knierim, U., Canali, E. and Winckler, C. 2009b. Reliability of measures of socio-positive and play behaviour in dairy and beef cattle. In Forkman, B. and Keeling, L. (Eds), *Assessment of Animal Welfare Measures for Dairy Cattle, Beef Cattle and Veal Calves*, pp. 175–188, Cardiff, UK: Cardiff University.

Lensink, B. J., van Reenen, C. G., Engel, B., Rodenburg, T. B. and Veissier, I. 2003. Repeatability and reliability of an approach test to determine calves' responsiveness to humans – A brief report. *Appl. Anim. Behav. Sci.* 83(4), 325–330.

Lidfors, L. and Isberg, L. 2003. Intersucking in dairy cattle – review and questionnaire. *Appl. Anim. Behav. Sci.* 80(3), 207–231.

Magnusson, B. and Örnemark, U. 2014. The fitness for purpose of analytical methods – A laboratory guide to method validation and related topics. *Eurachem*, 2nd edition, 62 pp., available at www.eurachem.org.

Magrin, L., Gottardo, F., Brscic, M., Contiero, B. and Cozzi, G. 2019. Health, behaviour and growth performance of Charolais and Limousin bulls fattened on different types of flooring. *Animal* 13(11), 2603–2611.

Mandel, R., Whay, H. R., Klement, E. and Nicol, C. J. 2016. Invited review: Environmental enrichment of dairy cows and calves in indoor housing. *J. Dairy Sci.* 99(3), 1695–1715.

Martin, P. and Bateson, P. 2007. *Measuring Behaviour*, Cambridge, UK: Cambridge University Press.

Metz, J. H. M., Dijkstra, T., Franken, P. and Frankena, K. 2015. Development and application of a protocol to evaluate herd welfare in Dutch dairy farms. *Livest. Sci.* 180, 183–193.

Plesch, G., Broerkens, N., Laister, S., Winckler, C. and Knierim, U. 2010. Reliability and feasibility of selected measures concerning resting behaviour for the on-farm welfare assessment in dairy cows. *Appl. Anim. Behav. Sci.* 126(1–2), 19–26.

Pritchard, J. C., Barr, A. R. S. and Whay, H. R. 2006. Validity of a behavioural measure of heat stress and a skin tent test for dehydration in working horses and donkeys. *Equine Vet. J.* 38(5), 433–438.

Reenen, K. and Engel, B. 2004. Validation of welfare measures. *Welfare Quality Meeting*, Helsinki.

Rushen, J., Chapina, N. and de Passillé, A.-M. 2012. Automated monitoring of behavioural-based animal welfare indicators. *Anim. Welf.* 21(3), 339–350.

Scott, E. M., Nolan, A. M. and Fitzpatrick, J. L. 2001. Conceptual and methodological issues related to welfare assessment: A framework for measurement. *Acta Agric. Scand. Sect. A Anim. Sci. Suppl.* 30, 5–10.

Sorensen, J. T., Rousing, T., Moller, S. H., Bonde, M. and Hegelund, L. 2007. On-farm welfare assessment systems: What are the recording costs? *Anim. Welf.* 16, 237-239.

Špinka, M. and Dembele, I. Panamá, J. and Stìhulová, I. 2005. Lame dairy cows have shorter avoidance distances. In 39th International Congress of the International Society for Applied Ethology, Sagamihara, Japan, p. 83.

Stafleu, F. R., Grommers, F. J. and Vorstenbosch, J. 1996. Animal welfare: evolution and erosion of a moral concept. *Anim. Welf.* 5, 225-234.

Taverniers, I., De Loose, M. and Van Bockstaele, E. 2004. Trends in quality in the analytical laboratory. II. Analytical method validation and quality assurance. *Trends Anal. Chem.* 23(8), 535-552.

Thompson, M., Ellison, S. L. R. and Wood, R. 2002. Harmonized guidelines for single-laboratory validation of methods of analysis - (IUPAC technical report). *Pure Appl. Chem.* 74(5), 835-855.

Tuyttens, F. A. M., de Graaf, S., Heerkens, J. L. T., Jacobs, L., Nalon, E., Ott, S., Stadig, L., Van Laer, E. and Ampe, B. 2014. Observer bias in animal behaviour research: Can we believe what we score, if we score what we believe? *Anim. Behav.* 90, 273-280.

Veissier, I. and Boissy, A. 2007. Stress and welfare: Two complementary concepts that are intrinsically related to the animal's point of view. *Physiol Behav.* 92, 429-433.

Vessman, J., Stefan, R. I., Van Staden, J. F., Danzer, K., Lindner, W., Burns, D. T., Fajgelj, A. and Müller, H. 2001. Selectivity in analytical chemistry (IUPAC recommendations 2001). *Pure Appl. Chem.* 73(8), 1381-1386.

Welfare Quality®. 2009. *Welfare Quality® Assessment Protocol for Cattle (Fattening Cattle, Dairy Cows, Veal Calves).* Welfare Quality® Consortium, Lelystad, The Netherlands.

Whay, H. R., Main, D. C., Green, L. E. and Webster, A. J. 2003. Assessment of the welfare of dairy cattle using animal-based measurements: Direct observations and investigation of farm records. *Vet. Rec.* 153(7), 197-202.

Winckler, C. 2018. Assessment of cattle welfare: Approaches, goals and next steps on farms. In Tucker, C. B. (Ed.), *Advances in Cattle Welfare*, pp. 55-70, Duxford, UK: Woodhead Publishing.

Winckler, C. 2019. Assessing animal welfare at the farm level: Do we care sufficiently about the individual? *Anim. Welf.* 28(1), 77-82.

Winckler, C., Brinkmann, J. and Glatz, J. 2007. Long-term consistency of selected animal-related welfare parameters in dairy farms. *Anim. Welf.* 16, 197-199.

Chapter 5

Advances in precision livestock farming techniques for monitoring dairy cattle welfare

Henk Hogeveen and Mariska van der Voort, Wageningen University and Research, The Netherlands

1 Introduction

Society is increasingly interested about animal well-being in food-producing animals (Winckler, 2019). As a consequence, more and more regulations are created by public (government) and private (partners in the supply chain) entities to ensure animal welfare. Dairy cattle welfare is no exception to this trend. Welfare assessment systems are needed in order to study animal welfare. These can be very intensive (animal observation) systems which are used for research to, for instance, assess the impact of management and farm structure (housing) on dairy cattle welfare. These studies form the basis for welfare regulations.

Besides the use for regulations, there is a demand for methods to assess the overall welfare of farm animals. First attempts to assess animal welfare were resource based, because assessing minimum resource requirements is generally easier than to evaluate the impact of the production conditions on animals (Mench, 2003). Resource-based assessment intends to warrant

http://dx.doi.org/10.19103/AS.2020.0084.06

the provision of the necessary environmental conditions for optimal animal welfare. Animal-based assessment methods that were developed, such as the Welfare Quality® assessment protocol (Botreau et al., 2007), provide a thorough assessment of the impact of the actual rearing conditions on welfare. Country-specific protocols, such as the National Dairy FARM (Farmers Assuring Responsible Management) Program in the USA and the AssureWel program in the UK, are also developed and propagated. Moreover, individual milk processing cooperatives/companies are implementing their own dairy cattle welfare programs, often in collaboration with large customers.

All of these programs are mostly resource-oriented. Farm assessment happens on paper as well as through regular (but not frequent) farm visits by an assessor. This makes the implementation of comprehensive animal welfare assessment protocols at a commercial level difficult, time-consuming and expensive. Precision livestock farming (PLF) can offer tools that enable a farmer to monitor animals automatically, and thus to also monitor their health and welfare (Berckmans, 2014; Halachmi et al., 2019). Sensors (e.g. cameras, microphones and accelerometers), images, sounds and movements, combined with state-of-the-art algorithms, can be used to automatically monitor animal behavior or other animal-based measures. New technological innovations are thus appearing, although their practical implementation for welfare assessment is still limited. PLF may be well suited to assess the welfare of individual animals. A number of publications have pointed out the potential use of PLF for welfare monitoring (Nadimi et al., 2012; Rutter, 2014; Shalloo et al., 2018; Werner et al., 2019).

In this chapter an overview will be provided on current knowledge and future potential of PLF technology to enhance dairy cattle welfare. The potential of PLF technology will be described linked to the Five Domains framework (Mellor, 2017). Therefore, first this framework will be shortly described, after which the potential of the use of PLF for each of the domains will be given. Finally, the need for data processing algorithms will be described and discussed.

2 The five domains model for welfare assessment

The Five Freedoms framework of the UK Farm Animal Welfare Council (FAWC) is one of the most widely used frameworks used in relation to animal welfare (Mellor, 2017). Since animal welfare contains many aspects, attempts to create a simple definition of good animal welfare typically share two main components: that animals are physically healthy and that they are largely free from mental suffering. However, these simplistic definitions hide the fact that a multitude of factors contribute to animal welfare, and this is arguably the main reason it is not possible to use one common, simple definition. In contrast, the Five Freedoms framework reflects the multitude of factors that impact animal welfare.

With the Five Freedoms framework, the strengths and weaknesses of animal husbandry systems can be judged (Webster, 2008). The Five Freedoms framework has and is extensively used as the basis for animal welfare programs (e.g. McCulloch, 2013). In order to address limitations of the Five Freedoms framework, but staying closely connected, the Five Domains framework was developed (Mellor, 2015, 2017). Fogarty et al. (2019) have used this framework to evaluate the potential of PLF to monitor the welfare of sheep.

The Five Domains framework (Table 1) consists of five areas, domains, in which welfare can be compromised or enhanced. Three domains - nutrition, environment and health - contain survival-related factors. One domain contains factors regarding the external situation and one domain contains an overarching assessment of the influence of domains 1–4 on the affective experience of the animal. Because of the structure of this framework, animal welfare features in the fifth domain (mental state) are linked to the other four domains. For instance, presence of disease or injury (features in the health domain) is associated with pain, debility and sickness, which are features in the fifth domain (mental state). Because of that association, in the following four paragraphs, the potential application of PLF systems to monitor features in these four domains (nutrition, environment, health and behavior) will be described including aspects linked to the fifth domain (mental state).

3 The nutrition domain

One of the first applications of PLF was the introduction of automated concentrate feeders. The development started already in the 1960s (Harshbarger et al., 1965), but started to become widely applied in the 1980s (Van Asseldonk et al., 1998). In confinement systems, the use of automated feeding ensures intake of sufficient feed and that way prevents hunger. Although the currently used water systems cannot be seen as PLF, they are developed and designed in such a way that water is always available, as long as there are enough water troughs. One aspect in which sensor systems may play a role is to ensure a proper functioning of the water and feed systems. Water flow systems are used in pig farming and are even used in PLF applications (e.g. Jensen et al., 2017).

A specific form of feeding automation is the use of automated calf feeders (Jorgensen et al., 2017; Medrano-Galarza et al., 2018). Combined with a weighing scale, this allows for decision support regarding calf feeding and calf rearing decisions (Cantor et al., 2020; Eslamizad et al., 2018). Typically of these automated feeding and drinking systems is that they were developed to either improve production efficiency or reduce the use of labor (automation). However, by performing those tasks, these systems also support animal welfare by ensuring a freedom of hunger and thirst. Moreover, when the decision algorithms behind these systems would take balance of diet into account, these

Table 1 Key features and examples of the extended five domains model for welfare assessment (Mellor, 2015) as used by (Fogarty et al., 2019)

Physical/functional domains							
Survival-related factors						Situation-related factors	
1 Nutrition		2 Environment		3 Health		4 Behavior	
Restrictions:	Opportunities:	Unavoidable/ imposed conditions:	Available conditions:	Presence of:	Presence of:	Exercise of 'agency' impeded by:	'Agency' exercised by:
Water or food intake	Drink enough water	Thermal extremes	Thermally tolerable	Disease	Appropriate body condition score	Choices markedly restricted	Available engaging choices
Food quality and variety	Eat a balanced/ varied diet	Close confinement	Space for free movement	Injury	Good fitness level	Constraints on anima-animal interaction	Bonding
Voluntary overeating		Unpredictable events	Normal environment variation	Functional impairment			Play
Affective experience domain							
5 Mental state							
Negative:	Positive:	Negative forms of discomfort:	Positive forms of comfort:	Negative:	Positive:	Negative:	Positive calmness:
Thirst	Quenching thirst	Thermal	Thermal	Pain	Comforts of good health and high functional capacity	Anger	Engaged
Hunger	Pleasure of different tastes/ smells	Respiratory	Respiratory	Debility		Boredom	Maternally rewarded
		Auditory	Auditory	Sickness		Helplessness	Playfulness

aspects of dairy cattle welfare may also be supported by systems that are in use anyway. For instance, by monitoring rumination and activity, potential negative effects of dietary changes may be detected (Giaretta et al., 2019).

One aspect about dairy cattle nutrition requires specific attention. That is the issue of roughage intake. In confinement systems, the total roughage intake for a group of cows is often known. However, individual intake of roughage is lacking and individual cows vary in their actual roughage intake. Individual cow roughage intake systems are available (Devir et al., 1996), but these are only used in experimental settings due to the complexity of the system and the price. Recently, using 3D accelerometers, it is shown that it is possible to get insight in roughage intake of dairy cattle (Carpinelli et al., 2019; Zambelis et al., 2019; Reiter et al., 2018) as well as the feeding (Byskov et al., 2015; Ruuska et al., 2016) and drinking behavior of cattle (Williams et al., 2019). Positioning sensor systems may add to the prediction of feeding behavior (Barker et al., 2018; Mattachini et al., 2016). Also in grazing systems, roughage intake has been studied with sensor systems (Jemila and Priyadharsini, 2018; Oudshoorn et al., 2013). Using neck-mounted 3D accelerometers, it is possible to estimate grass intake (Andriamandroso et al., 2017). These developments open possibilities to monitor the feed and water intake and ensure the freedom of hunger and thirst from the cow behavior's point of view.

4 The environment domain

Animal welfare aspects in the environment domain are often related to the design and layout of a barn and PLF systems do not play a large role in that. A proper barn design may provide enough space to cows to guarantee free movement. Or, in climates with a risk of heat stress, barn designs do have equipment to mitigate the effects of the high temperature and humidity. By keeping cows in well-designed barns, animal comfort may be guaranteed as good as possible. Moreover, by having measuring equipment (sensors) in the barn to monitor the barn climate, a farmer can monitor that climate and take action when needed.

However, more and more it is important to evaluate the actual well-being of cows rather than the conditions. For instance, climatic circumstances may lead to heat stress in dairy cattle. Recent work has been carried out to monitor the thermal status of dairy cattle using 3D accelerometers (Bar et al., 2019). Algorithms were developed to monitor the breathing of cows and monitoring 'heavy breathing', an indication of heat stress. Based on this indicator, cooling interventions could be undertaken.

The space for free movement is another important aspect in the environment domain. By measuring the movements of cows, it will be possible to monitor whether cows can move freely. Especially accelerometers are able to carry out that task (Martiskainen et al., 2009). Currently a number of PLF technologies are available for this type of monitoring (e.g. Borchers et al., 2016). The freedom of movement is also associated with the ease of walking. The construction and the cleanliness of the floor is an important aspect for ease of walking. Also, suboptimal mobility leads to impairment of the ease of walking. Suboptimal mobility is associated with claw problems (O'Connor et al., 2019) and does also affect the ease of walking. Suboptimal mobility can be monitored by the use of different types of PLF sensor systems (e.g. Barker et al., 2018; Beer et al., 2016; Nechanitzky et al., 2016; Pastell et al., 2008, 2009; Thorup et al., 2015; Van de Gucht et al., 2017). It seems that the performance of these systems, in terms of sensitivity and specificity is not yet good enough to manage individual cows, but most probably these types of systems can be used to monitor the ease of movement of the whole herd.

5 The health domain

By ensuring good dairy cattle health, we can safeguard them from pain, debility and sickness. Animal health is, therefore, an important component of animal welfare. The most important diseases with regard to animal welfare are the so-called production diseases: diseases that are associated with dairy production and which are often relatively highly prevalent. The most important of these diseases are mastitis, claw disorders and metabolic disorders.

Until now, PLF applications have especially been studied to monitor dairy cattle health (Rutten et al., 2013). The most studied disease is mastitis. Obviously because of the (economic) importance of mastitis (Hogeveen et al., 2019) as well as because of the introduction of automatic milking, where mastitis detection could not anymore be done by a milker. Therefore, sensor systems were needed to replace the visual observations of the milker (Hogeveen et al., 2010). Mostly these sensors utilized electrical conductivity but recently somatic cell count sensors are becoming available as well (Dalen et al., 2019; Deng et al., 2020).

Foot disorders, associated with suboptimal mobility, are costly (Bruijnis et al., 2010) and might be more prevalent than mastitis; the prevalence of suboptimal mobility was more than 35% in a recent study (O'Connor et al., 2019). Since foot disorders are associated with pain it is an important disease from a welfare point of view (Bruijnis et al., 2012). Although some work has been carried out on PLF systems to detect suboptimal mobility (see the previous paragraph), it seems that the development of PLF applications for suboptimal mobility does not have a priority. Potentially that is because that work is aimed at the detection of cows with suboptimal mobility, a task that farmers can also carry out visually. No work has been carried out so far on using PLF to monitor suboptimal mobility at the herd level, which could be an interesting welfare monitoring application.

Some work has been carried out on the development of PLF systems to detect cows with metabolic disorders, both subclinical and clinical. A number of predictors are available (LeBlanc et al., 2005; Overton et al., 2017). One of the potential predictors is rumination, a predictor that is used a number of times to detect metabolic disorders using PLF (Kroger et al., 2016; Molfino et al., 2017; Reiter et al., 2018; Schirmann et al., 2009), potentially in combination with activity monitoring (Stangaferro et al., 2016). Often this work is done using neck or ear-mounted 3D accelerometers but rumen bolus sensors (Benaissa et al., 2019; Hamilton et al., 2019; Sturm et al., 2020), in-line beta-hydroxybutyrate sensors (Nielsen et al., 2005), electronic nose (Mottram, 1997; Mottram et al., 1999), the milk fat-to-protein ratio (King et al., 2019), lying (Kaufman et al., 2016) or behavioral patterns (Wagner et al., 2020) are also used.

Interestingly, all PLF applications in the health domain that are described so far are aimed at detecting a suboptimal situation (disease) at the cow level. Detection of a suboptimal situation may be used to initiate an intervention at cow-level, for example, treatment. The uptake of these systems is not always high (Steeneveld and Hogeveen, 2015; Borchers and Bewley, 2015; Stone, 2020) which may be associated by either a lack of clear intervention options or by a lack of accuracy of the PLF system in comparison with the farmers' accuracy. Until now, no authors have been looking at the value of PLF systems in monitoring health at the herd level. This type of monitoring is also useful with regard to the monitoring of animal welfare. By monitoring disease occurrence over time, a farmer and other interested stakeholders have a real-time overview of the health status of cows on an individual farm.

The usefulness of PLF has also been studied regarding bovine respiratory disease. Most of the studies were aimed at respiratory disease in youngstock, using 3D accelerometers (Swartz et al., 2017) and sound recordings to measure coughing (Carpentier et al., 2018; Ferrari et al., 2010; Vandermeulen et al., 2016).

Besides monitoring the presence of the diseases, the negative aspect of the health domain, PLF systems may also be helpful in monitoring positive aspects such as the body condition score (e.g. Bercovich et al., 2013; Mullins et al., 2019), milk production level as expected (e.g. Adriaens et al., 2018; Jensen et al., 2016), absence of fever (Kou et al., 2017), feeding behavior (Gonzalez et al., 2008; Knauer et al., 2018; Svensson and Jensen, 2007) or the absence of any disease signs (Eckelkamp and Bewley, 2020; King and DeVries, 2018). It may even be possible to use routinely available milk shipment data to monitor disease at the herd level (Fall et al., 2018).

6 The behavior domain

As indicated earlier, sensor systems so-far have been developed and studied with regard to animal, disease and production management. For dairy farmers, there is a (financial) value in those aspects of animal management, while animal behavior, as such, does not contain financial value. However, animal behavior is an essential aspect of animal welfare. Animals should be able to perform their natural behavior to prevent anger, boredom and helplessness. Besides the often mentioned feeding, standing and lying behavior, which are essential behavioral aspects, also social behavior (interaction with other cows) is an important aspect in the behavior domain.

PLF systems may be very suitable to monitor animal behavior (Rushen et al., 2012). Especially 3D accelerometers are able to distinguish between several types of behavior (standing, eating, lying, walking). Since lying is an important welfare indicator and can also be used to evaluate the comfort of stalls within barns, a number of studies have focused on algorithms to evaluate lying time of cows (e.g. Kok et al., 2015; Robert et al., 2011). Some other studies have more generally studied the use of PLF systems to explore the time budgets of dairy cattle (e.g. Arcidiacono et al., 2017; Benaissa et al., 2019; Grinter et al., 2019; Jaeger et al., 2019; Tamura et al., 2019).

Adding cow location sensors to the accelerometer sensors does improve the precision of the cow behavior monitoring (Wang et al., 2019). No studies have been carried out on monitoring cattle interaction (except for work on estrus detection), but systems that do monitor the indoor position of cows can be an interesting tool to monitor social interaction between cows (Pastell et al., 2018). Another potential PLF system that may be used to monitor social interaction between animals is the use of systems such as underdevelopment for the Coronavirus contact-tracing apps working with mobile phone technology (Fraser et al., 2020). Such systems allow the evaluation of the number of cows one cow has been interacting with, potentially in combination with the place where that interaction happened. An example of utilizing the potential of monitoring the interaction between cows was given by Huzzey et al. (2014), who looked at social competition in front of automated feeding systems.

Grazing behavior is another aspect that pleases a role. PLF may be useful to enable efficient grazing systems by virtual fencing (Lomax et al., 2019). Just as with in-house feeding behavior, 3D accelerometers are also used to evaluate grazing behavior (Werner et al., 2019; Pereira et al., 2020; Gonzalez et al., 2015; Poulopoulou et al., 2019). Especially with this aspect, cow position information can be of additional value to monitor where cows have been grazing besides how cows have been grazing. From a regulatory perspective the monitoring of grazing may also be interesting since in some countries or with some dairy-processing companies, farmers are rewarded when their cows graze, which means that a system should be in place to assess grazing of cows.

7 The need for algorithms to monitor dairy cattle welfare

Since much of the research done so far was not aimed specifically at monitoring animal welfare, we chose to give an overview of the potential use of PLF in monitoring welfare and to not try to provide the accuracy with which monitoring dairy cattle welfare can be carried out. Nor did we try to be complete in all references or all PLF systems. Given the currently described PLF systems, a summary of the PLF systems that may play a role can be created:

- Accelerometers. These may be placed in the ear, around the leg or the neck. Their basic application is estrus detection and they are widely used in dairy farms. Depending on their position they can be used in monitor feeding and other behavior of dairy cattle and be useful to evaluate a wide range of features relating to animal welfare.
- Cow-positioning electronics. These may be placed in the ear, around the leg or in the neck, often in combination with accelerometers. Cow-positioning sensors are in practice and are used to find specific cows, but can be useful to evaluate social interaction between cows and some types of behavior.
- Temperature sensors. These may be placed in the ear, in the milk line and in the rumen. Temperature measurements can be used in monitoring animal health and heat stress.
- pH sensors. This sensor may be placed in the rumen and can be used to monitor metabolic status of cows, aimed at detecting subclinical acidosis, or deviations in the feeding ration.
- Cameras. A number of different cameras may be utilized in monitoring animal welfare. In an overview placement, cameras can be used to monitor cows' positions as well as social interaction and discomfort. Closer to the cow they can be used to evaluate body condition score and thus the general well-being of a cow. In another position, cameras may be used to monitor mobility of cows. Finally, thermal cameras may be used to monitor cows' health.
- Sensors measuring parameters in milk. A wide range of sensors are available, mostly aimed at monitoring udder health and milk production (milk yield, fat and protein percentage). The production parameters can be used to monitor the absence of disease.

Although all of the above sensor systems have been described in literature and have shown to be potentially useful for a specific aspect of dairy farm management, specific algorithms should be developed to enable the use of these sensors in animal welfare monitoring. The developments in machine learning and big data analytics have been extensive in the past years (Morota et al., 2018). It is expected that the machine-learning toolbox is sufficiently filled to create proper algorithms to monitor dairy cattle welfare based upon PLF systems. However, before useful algorithms can be developed, the performance needs for those algorithms should be known. Until now, most machine-learning algorithms are aimed at the detection of cases of disease in individual animals. Depending on the type of disease, the prevalence of disease, the potential welfare and production consequences of missing a disease and the costs for intervention, specifications for the detection performance may be defined. Test characteristics typically consist of the sensitivity, the specificity and the time window in which the disease should be detected. However, until now, minimal test characteristics have only been provided for one disease: clinical mastitis (Hogeveen et al., 2010).

For animal welfare issues, interventions are most probably provided at the herd level, rather than at the individual cow level. Therefore, the individual cow assessment of animal welfare features need to be aggregated. Performance of algorithms to monitor dairy cattle welfare at the herd level should be evaluated at the herd level and not at the individual cow level. Imperfections at the cow level algorithms to assess welfare, if random, may be less important when individual cow assessments are aggregated at the herd level.

Besides the use of PLF for production monitoring and optimization and for animal disease management, PLF may also be used to assess animal welfare on dairy farms. Such an assessment may be used by the farm management to intervene. Farm managers should receive an alert when dairy cattle behavior, for instance, lying behavior is changing. These alerts could be linked to key performance indicators (KPIs) related to animal welfare. Such KPIs can consist of the proportion of cows with abnormal behavior, or the proportion of cows potentially suffering from a negative situation, such as heat stress, pain, hunger or thirst (aspects from the fifth domain, the mental state). Linked to these KPIs, algorithms should be provided that provide alerts. The needed performance for such alerts depends on the severity of the feature. Issues with hunger/thirst or severe pain should be alerted very quickly and should have a high sensitivity. Issues regarding changes in cow behavior may last longer. Issues where quick intervention is possible may need different performance than issues that need further diagnostics. Typically algorithms that may be used for such tasks could be based on finding a deviation in the trend, such as dynamic linear programming (e.g. Cornou et al., 2011, 2014) or statistical process control (e.g. Huybrechts et al., 2014). But before algorithms can be trained, it would be good that experts in the fields of animal welfare, farm management, animal health and data management would get together to work out the demands for algorithms given the available PLF systems. The scientific knowledge so far is not sufficient in that respect. As an example, for mastitis management a number of experts have been together and worked out demands for PLF systems to management of mastitis at various levels, beyond the detection of clinical mastitis (Hogeveen et al., 2020).

8 Conclusion

At the moment, the availability of more and more, relatively cheap technology in combination with the processing power of modern computers is bringing a large change to the management of dairy farms (Cabrera et al., 2020; Barkema et al., 2015). Precision livestock farming has brought a number of sensor and automated data acquisition technologies to commercial farms (Berckmans, 2014). Most of these sensor and data acquisition systems are primarily developed to improve production efficiency. As long as precision livestock technology is not associated with premium prices, financial efficiency of these systems needs to be achieved through improvements in resource-use efficiency. However, many of the data acquired by precision technology are pertinent to welfare assessment (Rutter, 2014) and might be a basis for premium prices. Moreover, by linking the precision technology to adequate animal-based decision support systems, optimal use of (environmental) resources may be supported (Shalloo et al., 2018), improving the sustainability of livestock production.

There is an increasing societal demand for animal welfare assessment (Winckler, 2019). Over time, processors and/or retailers may demand that all livestock products come with objective, sensor-derived farm data to demonstrate they were produced to acceptable welfare standards. Current resource-based assessment systems do have a value in enabling a systematic evaluation of animal welfare. However, it has been criticized for a number of reasons: (1) using too few animal-based measures, (2) time-consuming yet infrequent assessments and (3) inflexible weighting of different welfare factors (McCulloch, 2013). Besides the inflexibility of the current weighting of the criteria, they do not necessarily conform to the public's perceptions of the relative importance of the factors that contribute to animal welfare. For example, consumers seem to accept quite a number of negative animal welfare aspects (e.g. disease) as long as the animals are free ranged (Bergstra et al., 2017). In other words, animal welfare assessment systems should be based on a flexible system of weighting the various aspects of animal welfare, in order to meet consumers' preferences as well as scientific knowledge. Moreover, in order to be useful, assessment of animal welfare should be animal-based and easy to perform in order to be used. PLF systems may, as is shown in this chapter, be of good help in obtaining and monitoring an overall view of dairy cattle welfare.

Currently, the uptake of precision technologies available varies between different livestock systems and European regions. Although the greatest uptake of PLF technologies has, to date, been in intensively managed livestock production systems with much attention for individual animals (e.g. more than 15% of Dutch dairy farmers have 3D accelerometers, Steeneveld and Hogeveen, 2015), there is no fundamental reason why they could not also be used in more extensively managed systems (e.g. Gonzalez et al., 2015; Rutter, 2014). The benefit of PLF systems to monitor animal welfare is only partly for a dairy farmer. That part is linked to the use of PLF systems to make production more efficient and to reduce the negative consequences of animal disease. Although farmers have an intrinsic interest in animal welfare (Valeeva et al., 2007), they will most likely not invest much in sensor systems to monitor animal welfare. By utilizing PLF systems that are used by farmers for other purposes (e.g. 3D accelerometers which are very cost-efficient to be used for estrus detection (Rutten et al., 2013), the animal welfare monitoring can be carried out without any additional investments in hardware. Typically other stakeholders in the dairy value chain may be interested in utilizing such PLF-based welfare monitoring systems, for instance, for milk quality certification programs. Business models to pay for the infrastructure and the additional software should be developed. Finally, we should also note that, from an ethical point of view there are some remarks on the use of PLF systems, because they 'objectify' the animal (Bos et al., 2018).

In conclusion, based on the Five Domains framework, an overview has been given on the potential use of PLF systems to monitor animal welfare. Currently, sensor technology is available but has hardly been used to monitor animal welfare. We are at the start of these developments. Much is still needed in defining the needs for animal welfare monitoring. That work should provide the foundation on which algorithms can be created that utilize the PLF technology data for monitoring animal welfare.

9 Where to look for further information

In this chapter, we provided an overview of the precision livestock techniques that may be used to monitor animal welfare. We showed that with the currently available techniques it is possible to monitor animal welfare. Yet, the scientific literature on this topic is highly fragmented, the association between single animal welfare aspects and data collected by sensor systems is provided, often from the point of view of animal disease management (disease detection) or measuring production efficiency (looking at feed intake and feeding behavior). Moreover, studies are either carried out from a more technological point of view (for example aimed at improved sensing technology or algorithms) or from a more animal welfare point of view. A first step for future research would be the development of a concise animal welfare monitoring framework from the point of view of precision livestock technology. Such a framework should be developed by a trans-disciplinary group of researchers representing a number of research and implementation disciplines. Within such a framework researchers, and within their own discipline may work out elements. The information brought together in this chapter can be used as a starting point.

10 References

Adriaens, I., Huybrechts, T., Aernouts, B., Geerinckx, K., Piepers, S., De Ketelaere, B. and Saeys, W. 2018. Method for short-term prediction of milk yield at the quarter level to improve udder health monitoring. *Journal of Dairy Science* 101(11), 10327–10336.

Andriamandroso, A. L. H., Lebeau, F., Beckers, Y., Froidmont, E., Dufrasne, I., Heinesch, B., Dumortier, P., Blanchy, G., Blaise, Y. and Bindelle, J. 2017. Development of an open-source algorithm based on inertial measurement units (IMU) of a smartphone to detect cattle grass intake and ruminating behaviors. *Computers and Electronics in Agriculture* 139, 126–137.

Arcidiacono, C., Porto, S. M. C., Mancino, M. and Cascone, G. 2017. Development of a threshold-based classifier for real-time recognition of cow feeding and standing behavioural activities from accelerometer data. *Computers and Electronics in Agriculture* 134, 124–134.

Bar, D., Kaim, M., Flamenbaum, I., Hanochi, B. and Toaff-Rosenstein, R. L. 2019. Technical note: accelerometer-based recording of heavy breathing in lactating and dry cows as an automated measure of heat load. *Journal of Dairy Science* 102(4), 3480–3486.

Barkema, H. W., Von Keyserlingk, M. A. G., Kastelic, J. P., Lam, T. J., Luby, C., Roy, J. P., LeBlanc, S. J., Keefe, G. P. and Kelton, D. F. 2015. Invited review: changes in the dairy industry affecting dairy cattle health and welfare. *Journal of Dairy Science* 98(11), 7426–7445.

Barker, Z. E., Diosdado, J. A. V., Codling, E. A., Bell, N. J., Hodges, H. R., Croft, D. P. and Amory, J. R. 2018. Use of novel sensors combining local positioning and acceleration to measure feeding behavior differences associated with lameness in dairy cattle. *Journal of Dairy Science* 101(7), 6310–6321.

Beer, G., Alsaaod, M., Starke, A., Schuepbach-Regula, G., Muller, H., Kohler, P. and Steiner, A. 2016. Use of extended characteristics of locomotion and feeding behavior for automated identification of lame dairy cows. *PLoS ONE* 11(5), e0155796.

Benaissa, S., Tuyttens, F. A. M., Plets, D., Cattrysse, H., Martens, L., Vandaele, L., Joseph, W. and Sonck, B. 2019. Classification of ingestive-related cow behaviours using RumiWatch halter and neck-mounted accelerometers. *Applied Animal Behaviour Science* 211, 9–16.

Berckmans, D. 2014. Precision livestock farming technologies for welfare management in intensive livestock systems. *Revue Scientifique et Technique* 33(1), 189–196.

Bercovich, A., Edan, Y., Alchanatis, V., Moallem, U., Parmet, Y., Honig, H., Maltz, E., Antler, A. and Halachmi, I. 2013. Development of an automatic cow body condition scoring using body shape signature and fourier descriptors. *Journal of Dairy Science* 96(12), 8047–8059.

Bergstra, T., Hogeveen, H., Kuiper, W. E., Oude Lansink, A. G. J. M. and Stassen, E. N. 2017. Attitudes of Dutch citizens toward sow husbandry with regard to animals, humans, and the environment. *Anthrozoös* 30(2), 195–211.

Borchers, M. R. and Bewley, J. M. 2015. An assessment of producer precision dairy farming technology use, prepurchase considerations, and usefulness. *Journal of Dairy Science* 98(6), 4198–4205.

Borchers, M. R., Chang, Y. M., Tsai, I. C., Wadsworth, B. A. and Bewley, J. M. 2016. A validation of technologies monitoring dairy cow feeding, ruminating, and lying behaviors. *Journal of Dairy Science* 99(9), 7458–7466.

Bos, J. M., Bovenkerk, B., Feindt, P. H. and Van Dam, Y. K. 2018. The quantified animal: precision livestock farming and the ethical implications of objectification. *Food Ethics* 2(1), 77–92.

Botreau, R., Bonde, M., Butterworth, A., Perny, P., Bracke, M. B. M., Capdeville, J. and Veissier, I. 2007. Aggregation of measures to produce an overall assessment of animal welfare. Part 1: a review of existing methods. *Animal: An International Journal of Animal Bioscience* 1(8), 1179–1187.

Bruijnis, M. R. N., Beerda, B., Hogeveen, H. and Stassen, E. N. 2012. Assessing the welfare impact of foot disorders in dairy cattle by a modeling approach. *Animal: An International Journal of Animal Bioscience* 6(6), 962–970.

Bruijnis, M. R. N., Hogeveen, H. and Stassen, E. N. 2010. Assessing economic consequences of foot disorders in dairy cattle using a dynamic stochastic simulation model. *Journal of Dairy Science* 93(6), 2419–2432.

Byskov, M. V., Nadeau, E., Johansson, B. E. O. and Norgaard, P. 2015. Variations in automatically recorded rumination time as explained by variations in intake of dietary fractions and milk production, and between-cow variation. *Journal of Dairy Science* 98(6), 3926–3937.

Cabrera, V. E., Barrientos-Blanco, J. A., Delgado, H. and Fadul-Pacheco, L. 2020. Symposium review: real-time continuous decision making using big data on dairy farms. *Journal of Dairy Science* 103(4), 3856–3866.

Cantor, M. C., Pertuisel, C. H. and Costa, J. H. C. 2020. Technical note: estimating body weight of dairy calves with a partial-weight scale attached to an automated milk feeder. *Journal of Dairy Science* 103(2), 1914–1919.

Carpentier, L., Berckmans, D., Youssef, A., Berckmans, D., Van Waterschoot, T., Johnston, D., Ferguson, N., Earley, B., Fontana, I., Tullo, E., Guarino, M., Vranken, E. and Norton, T. 2018. Automatic cough detection for bovine respiratory disease in a calf house. *Biosystems Engineering* 173, 45–56.

Carpinelli, N. A., Rosa, F., Grazziotin, R. C. B. and Osorio, J. S. 2019. Technical note: a novel approach to estimate dry matter intake of lactating dairy cows through multiple on-cow accelerometers. *Journal of Dairy Science* 102(12), 11483–11490.

Cornou, C., Lundbye-Christensen, S. and Kristensen, A. R. 2011. Modelling and monitoring sows' activity types in farrowing house using acceleration data. *Computers and Electronics in Agriculture* 76(2), 316–324.

Cornou, C., Ostergaard, S., Ancker, M. L., Nielsen, J. and Kristensen, A. R. 2014. Dynamic monitoring of reproduction records for dairy cattle. *Computers and Electronics in Agriculture* 109, 191–194.

Dalen, G., Rachah, A., Nørstebø, H., Schukken, Y. H. and Reksen, O. 2019. The detection of intramammary infections using online somatic cell counts. *Journal of Dairy Science* 102(6), 5419–5429.

Deng, Z. J., Hogeveen, H., Lam, T. J. G. M., Van Der Tol, R. and Koop, G. 2020. Performance of online somatic cell count estimation in automatic milking systems. *Frontiers in Veterinary Science* 7, 221.

Devir, S., Hogeveen, H., Hogewerf, P. H., Ipema, A. H., Ketelaardelauwere, C. C. K., Rossing, W., Smits, A. C. and Stefanowska, J. 1996. Design and implementation of a system for automatic milking and feeding. *Canadian Agricultural Engineering* 38, 107–113.

Eckelkamp, E. A. and Bewley, J. M. 2020. On-farm use of disease alerts generated by precision dairy technology. *Journal of Dairy Science* 103(2), 1566–1582.

Eslamizad, M., Tummler, L. M., Derno, M., Hoch, M. and Kuhla, B. 2018. Technical note: development of a pressure sensor-based system for measuring rumination time in pre-weaned dairy calves. *Journal of Animal Science* 96(11), 4483–4489.

Fall, N., Ohlson, A., Emanuelson, U. and Dohoo, I. 2018. Exploring milk shipment data for their potential for disease monitoring and for assessing resilience in dairy farms. *Preventive Veterinary Medicine* 154, 23–28.

Ferrari, S., Piccinini, R., Silva, M., Exadaktylos, V., Berckmans, D. and Guarino, M. 2010. Cough sound description in relation to respiratory diseases in dairy calves. *Preventive Veterinary Medicine* 96(3–4), 276–280.

Fogarty, E. S., Swain, D. L., Cronin, G. M. and Trotter, M. 2019. A systematic review of the potential uses of on-animal sensors to monitor the welfare of sheep evaluated using the five domains model as a framework. *Animal Welfare* 28(4), 407–420.

Fraser, C., Abeler-Doerner, L., Ferretti, L., Parker, M., Kendall, M. and Bonsall, D. 2020. *Digital Contact Tracing: Comparing the Capabilities of Centralised and Decentralised Data Architectures to Effectively Suppress the COVID-19 Epidemic Whilst Maximising Freedom of Movement and Maintaining Privacy*, Nature Publishing, preprint https://github.com/BDI-pathogens/covid-19_instant_tracing/blob/master/Centralised%20and%20decentralised%20systems%20for%20contact%20tracing.pdf.

Giaretta, E., Mordenti, A. L., Canestrari, G., Palmonari, A. and Formigoni, A. 2019. Automatically monitoring of dietary effects on rumination and activity of finishing heifers. *Animal Production Science* 59(10), 1931–1940.

Gonzalez, L. A., Bishop-Hurley, G. J., Handcock, R. N. and Crossman, C. 2015. Behavioral classification of data from collars containing motion sensors in grazing cattle. *Computers and Electronics in Agriculture* 110, 91–102.

Gonzalez, L. A., Tolkamp, B. J., Coffey, M. P., Ferret, A. and Kyriazakis, I. 2008. Changes in feeding behavior as possible indicators for the automatic monitoring of health disorders in dairy cows. *Journal of Dairy Science* 91(3), 1017–1028.

Grinter, L. N., Campler, M. R. and Costa, J. H. C. 2019. Technical note: validation of a behavior-monitoring collar's precision and accuracy to measure rumination, feeding, and resting time of lactating dairy cows. *Journal of Dairy Science* 102(4), 3487–3494.

Halachmi, I., Guarino, M., Bewley, J. and Pastell, M. 2019. Smart animal agriculture: application of real-time sensors to improve animal well-being and production. *Annual Review of Animal Biosciences* 7, 403–425.

Hamilton, A. W., Davison, C., Tachtatzis, C., Andonovic, I., Michie, C., Ferguson, H. J., Somerville, L. and Jonsson, N. N. 2019. Identification of the rumination in cattle using support vector machines with motion-sensitive bolus sensors. *Sensors* 19(5), 1165.

Harshbarger, K. E., Olver, E. F. and Shove, G. C. 1965. Effects of an automatic water-concentrate feeder on milk production. *Journal of Dairy Science* 48, 794.

Hogeveen, H., Kamphuis, C., Steeneveld, W. and Mollenhorst, H. 2010. Sensors and clinical mastitis-the Quest for the perfect alert. *Sensors* 10(9), 7991–8009.

Hogeveen, H., Klaas, I. C., Dalen, G., Honig, H., Zecconi, A., Kelton, D. F. and Sanchez Mainar, M. 2020. Novel approaches to manage mastitis support by sensor systems. Submitted for publication.

Hogeveen, H., Steeneveld, W. and Wolf, C. A. 2019. Production diseases reduce the efficiency of dairy production: a review of the results, methods, and approaches

Regarding the economics of mastitis. *Annual Review of Resource Economics* 11(1), 289-312.

Huybrechts, T., Mertens, K., De Baerdemaeker, J., De Ketelaere, B. and Saeys, W. 2014. Early warnings from automatic milk yield monitoring with online synergistic control. *Journal of Dairy Science* 97(6), 3371-3381.

Huzzey, J. M., Weary, D. M., Tiau, B. Y. F. and Von Keyserlingk, M. A. G. 2014. Short communication: automatic detection of social competition using an electronic feeding system. *Journal of Dairy Science* 97(5), 2953-2958.

Jaeger, M., Brugemann, K., Brandt, H. and Konig, S. 2019. Associations between precision sensor data with productivity, health and welfare indicator traits in native black and white dual-purpose cattle under grazing conditions. *Applied Animal Behaviour Science* 212, 9-18.

Jemila, J. S. and Priyadharsini, S. S. 2018. A sensor-based forage monitoring of grazing cattle in dairy farming. *International Journal on Smart Sensing and Intelligent Systems* 11(1), 1-9.

Jensen, D. B., Hogeveen, H. and De Vries, A. 2016. Bayesian integration of sensor information and a multivariate dynamic linear model for prediction of dairy cow mastitis. *Journal of Dairy Science* 99(9), 7344-7361.

Jensen, D. B., Toft, N. and Kristensen, A. R. 2017. A multivariate dynamic linear model for early warnings of diarrhea and pen fouling in slaughter pigs. *Computers and Electronics in Agriculture* 135, 51-62.

Jorgensen, M. W., Adams-Progar, A., De Passille, A. M., Rushen, J., Godden, S. M., Chester-Jones, H. and Endres, M. I. 2017. Factors associated with dairy calf health in automated feeding systems in the Upper Midwest United States. *Journal of Dairy Science* 100(7), 5675-5686.

Kaufman, E. I., LeBlanc, S. J., Mcbride, B. W., Duffield, T. F. and Devries, T. J. 2016. Short communication: association of lying behavior and subclinical ketosis in transition dairy cows. *Journal of Dairy Science* 99(9), 7473-7480.

King, M. T. M. and DeVries, T. J. 2018. Graduate Student Literature Review: detecting health disorders using data from automatic milking systems and associated technologies. *Journal of Dairy Science* 101(9), 8605-8614.

King, M. T. M., Duffield, T. F. and Devries, T. J. 2019. Short communication: assessing the accuracy of inline milk fat-to-protein ratio data as an indicator of hyperketonemia in dairy cows in herds with automated milking systems. *Journal of Dairy Science* 102(9), 8417-8422.

Knauer, W. A., Godden, S. M., Dietrich, A., Hawkins, D. M. and James, R. E. 2018. Evaluation of applying statistical process control techniques to daily average feeding behaviors to detect disease in automatically fed group-housed preweaned dairy calves. *Journal of Dairy Science* 101(9), 8135-8145.

Kok, A., Van Knegsel, A. T. M., Van Middelaar, C. E., Hogeveen, H., Kemp, B. and De Boer, I. J. M. 2015. Technical note: validation of sensor-recorded lying bouts in lactating dairy cows using a 2-sensor approach. *Journal of Dairy Science* 98(11), 7911-7916.

Kou, H. X., Zhao, Y. Q., Ren, K., Chen, X. L., Lu, Y. Q. and Wang, D. 2017. Automated measurement of cattle surface temperature and its correlation with rectal temperature. *PLoS ONE* 12(4), e0175377.

Kroger, I., Humer, E., Neubauer, V., Kraft, N., Ertl, P. and Zebeli, Q. 2016. Validation of a noseband sensor system for monitoring ruminating activity in cows under different feeding regimens. *Livestock Science* 193, 118-122.

LeBlanc, S. J., Leslie, K. E. and Duffield, T. F. 2005. Metabolic predictors of displaced abomasum in dairy cattle. *Journal of Dairy Science* 88(1), 159-170.

Lomax, S., Colusso, P. and Clark, C. E. F. 2019. Does virtual fencing work for grazing dairy cattle? *Animals: An Open Access Journal from MDPI* 9(7), 429.

Martiskainen, P., Jarvinen, M., Skon, J. P., Tiirikainen, J., Kolehmainen, M. and Mononen, J. 2009. Cow behaviour pattern recognition using a three-dimensional accelerometer and support vector machines. *Applied Animal Behaviour Science* 119(1-2), 32-38.

Mattachini, G., Riva, E., Perazzolo, F., Naldi, E. and Provolo, G. 2016. Monitoring feeding behaviour of dairy cows using accelerometers. *Journal of Agricultural Engineering* 47(1), 54-58.

McCulloch, S. P. 2013. A critique of FAWC's five freedoms as a framework for the analysis of animal welfare. *Journal of Agricultural and Environmental Ethics* 26(5), 959-975.

Medrano-Galarza, C., LeBlanc, S. J., Devries, T. J., Jones-Bitton, A., Rushen, J., De Passille, A. M., Endres, M. I. and Haley, D. B. 2018. Effect of age of introduction to an automated milk feeder on calf learning and performance and labor requirements. *Journal of Dairy Science* 101(10), 9371-9384.

Mellor, D. J. 2015. Positive animal welfare states and reference standards for welfare assessment. *New Zealand Veterinary Journal* 63(1), 17-23.

Mellor, D. J. 2017. Operational details of the five domains model and its key applications to the assessment and management of animal welfare. *Animals: An Open Access Journal from MDPI* 7(8), 60.

Mench, J. A. 2003. Assessing animal welfare at the farm and group level: a United States perspective. *Animal Welfare* 12, 493-503.

Molfino, J., Clark, C. E. F., Kerrisk, K. L. and Garcia, S. C. 2017. Evaluation of an activity and rumination monitor in dairy cattle grazing two types of forages. *Animal Production Science* 57(7), 1557-1562.

Morota, G., Ventura, R. V., Silva, F. F., Koyama, M. and Fernando, S. C. 2018. Big data analytics and precision animal agriculture symposium: machine learning and data mining advance predictive big data analysis in precision animal agriculture. *Journal of Animal Science* 96(4), 1540-1550.

Mottram, T. 1997. Automatic monitoring of the health and metabolic status of dairy cows. *Livestock Production Science* 48(3), 209-217.

Mottram, T. T., Dobbelaar, P., Schukken, Y. H., Hobbs, P. J. and Bartlett, P. N. 1999. An experiment to determine the feasibility of automatically detecting hyperketonaemia in dairy cows. *Livestock Production Science* 61(1), 7-11.

Mullins, I. L., Truman, C. M., Campler, M. R., Bewley, J. M. and Costa, J. H. C. 2019. Validation of a commercial automated body condition scoring system on a commercial dairy farm. *Animals: An Open Access Journal from MDPI* 9(6), 287.

Nadimi, E. S., Jorgensen, R. N., Blanes-Vidal, V. and Christensen, S. 2012. Monitoring and classifying animal behavior using ZigBee-based mobile ad hoc wireless sensor networks and artificial neural networks. *Computers and Electronics in Agriculture* 82, 44-54.

Nechanitzky, K., Starke, A., Vidondo, B., Muller, H., Reckardt, M., Friedli, K. and Steiner, A. 2016. Analysis of behavioral changes in dairy cows associated with claw horn lesions. *Journal of Dairy Science* 99(4), 2904-2914.

Nielsen, N. I., Friggens, N. C., Chagunda, M. G. G. and Ingvartsen, K. L. 2005. Predicting risk of ketosis in dairy cows using in-line measurements of beta-hydroxybutyrate: a biological model. *Journal of Dairy Science* 88(7), 2441-2453.

O'Connor, A. H., Bokkers, E. A. M., De Boer, I. J. M., Hogeveen, H., Sayers, R., Byrne, N., Ruelle, E. and Shalloo, L. 2019. Associating cow characteristics with mobility scores in pasture-based dairy cows. *Journal of Dairy Science* 102(9), 8332–8342.

Oudshoorn, F. W., Cornou, C., Hellwing, A. L. F., Hansen, H. H., Munksgaard, L., Lund, P. and Kristensen, T. 2013. Estimation of grass intake on pasture for dairy cows using tightly and loosely mounted di- and tri-axial accelerometers combined with bite count. *Computers and Electronics in Agriculture* 99, 227–235.

Overton, T. R., Mcart, J. A. A. and Nydam, D. V. 2017. A 100-year Review: metabolic health indicators and management of dairy cattle. *Journal of Dairy Science* 100(12), 10398–10417.

Pastell, M., Frondelius, L., Jarvinen, M. and Backman, J. 2018. Filtering methods to improve the accuracy of indoor positioning data for dairy cows. *Biosystems Engineering* 169, 22–31.

Pastell, M., Hautala, M., Poikalainen, V., Praks, J., Veermae, I., Kujala, M. and Ahokas, J. 2008. Automatic observation of cow leg health using load sensors. *Computers and Electronics in Agriculture* 62(1), 48–53.

Pastell, M., Tiusanen, J., Hakojarvi, M. and Hanninen, L. 2009. A wireless accelerometer system with wavelet analysis for assessing lameness in cattle. *Biosystems Engineering* 104(4), 545–551.

Pereira, G. M., Heins, B. J., O'Brien, B., Mcdonagh, A., Lidauer, L. and Kickinger, F. 2020. Validation of an ear tag-based accelerometer system for detecting grazing behaviour of dairy cows. *Journal of Dairy Science* 103(4), 3529–3544.

Poulopoulou, I., Lambertz, C. and Gauly, M. 2019. Are automated sensors a reliable tool to estimate behavioural activities in grazing beef cattle? *Applied Animal Behaviour Science* 216, 1–5.

Reiter, S., Sattlecker, G., Lidauer, L., Kickinger, F., Ohlschuster, M., Auer, W., Schweinzer, V., Klein-Jobstl, D., Drillich, M. and Iwersen, M. 2018. Evaluation of an ear-tag-based accelerometer for monitoring rumination in dairy cows. *Journal of Dairy Science* 101(4), 3398–3411.

Robert, B. D., White, B. J., Renter, D. G. and Larson, R. L. 2011. Determination of lying behavior patterns in healthy beef cattle by use of wireless accelerometers. *American Journal of Veterinary Research* 72(4), 467–473.

Rushen, J., Chapinal, N. and De Passille, A. M. 2012. Automated monitoring of behavioural-based animal welfare indicators. *Animal Welfare* 21(3), 339–350.

Rutten, C. J., Velthuis, A. G. J., Steeneveld, W. and Hogeveen, H. 2013. Invited review: sensors to support health management on dairy farms. *Journal of Dairy Science* 96(4), 1928–1952.

Rutter, S. M. 2014. Smart technologies for detecting animal welfare status and delivering health remedies for rangeland systems. *Revue Scientifique et Technique* 33(1), 181–187.

Ruuska, S., Kajava, S., Mughal, M., Zehner, N. and Mononen, J. 2016. Validation of a pressure sensor-based system for measuring eating, rumination and drinking behaviour of dairy cattle. *Applied Animal Behaviour Science* 174, 19–23.

Schirmann, K., Von Keyserlingk, M. A. G., Weary, D. M., Veira, D. M. and Heuwieser, W. 2009. Technical note: validation of a system for monitoring rumination in dairy cows. *Journal of Dairy Science* 92(12), 6052–6055.

Shalloo, L., O' Donovan, M., Leso, L., Werner, J., Ruelle, E., Geoghegan, A., Delaby, L. and O'Leary, N. 2018. Review: grass-based dairy systems, data and precision technologies. *Animal: An International Journal of Animal Bioscience* 12(s2), s262–s271.

Stangaferro, M. L., Wijma, R., Caixeta, L. S., Al-Abri, M. A. and Giordano, J. O. 2016. Use of rumination and activity monitoring for the identification of dairy cows with health disorders: Part I. Metabolic and digestive disorders. *Journal of Dairy Science* 99(9), 7395–7410.

Steeneveld, W. and Hogeveen, H. 2015. Characterization of Dutch dairy farms using sensor systems for cow management. *Journal of Dairy Science* 98(1), 709–717.

Stone, A. E. 2020. Symposium review: the most important factors affecting adoption of precision dairy monitoring technologies. *Journal of Dairy Science* 103(6), 5740–5745.

Sturm, V., Efrosinin, D., Ohlschuster, M., Gusterer, E., Drillich, M. and Iwersen, M. 2020. Combination of sensor data and health monitoring for early detection of subclinical ketosis in dairy cows. *Sensors* 20(5), 1484.

Svensson, C. and Jensen, M. B. 2007. Short communication: identification of diseased calves by use of data from automatic milk feeders. *Journal of Dairy Science* 90(2), 994–997.

Swartz, T. H., Findlay, A. N. and Petersson-Wolfe, C. S. 2017. Short communication: automated detection of behavioral changes from respiratory disease in pre-weaned calves. *Journal of Dairy Science* 100(11), 9273–9278.

Tamura, T., Okubo, Y., Deguchi, Y., Koshikawa, S., Takahashi, M., Chida, Y. and Okada, K. 2019. Dairy cattle behavior classifications based on decision tree learning using 3-axis neck-mounted accelerometers. *Animal Science Journal = Nihon Chikusan Gakkaiho* 90(4), 589–596.

Thorup, V. M., Munksgaard, L., Robert, P. E., Erhard, H. W., Thomsen, P. T. and Friggens, N. C. 2015. Lameness detection via leg-mounted accelerometers on dairy cows on four commercial farms. *Animal: An International Journal of Animal Bioscience* 9(10), 1704–1712.

Valeeva, N. I., Lam, T. J. G. M. and Hogeveen, H. 2007. Motivation of dairy farmers to improve mastitis management. *Journal of Dairy Science* 90(9), 4466–4477.

Van Asseldonk, M. A., Huirne, R. B. M., Dijkhuizen, A. A., Tomaszewski, M. A. and Harbers, A. G. F. 1998. Effects of information technology on dairy farms in the Netherlands: an empirical analysis of milk production records. *Journal of Dairy Science* 81(10), 2752–2759.

Van de Gucht, T., Saeys, W., Van Weyenberg, S., Lauwers, L., Mertens, K., Vandaele, L., Vangeyte, J. and Van Nuffel, A. 2017. Automatic cow lameness detection with a pressure mat: effects of mat CD length and sensor resolution. *Computers and Electronics in Agriculture* 134, 172–180.

Vandermeulen, J., Bahr, C., Johnston, D., Earley, B., Tullo, E., Fontana, I., Guarino, M., Exadaktylos, V. and Berckmans, D. 2016. Early recognition of bovine respiratory disease in calves using automated continuous monitoring of cough sounds. *Computers and Electronics in Agriculture* 129, 15–26.

Wagner, N., Antoine, V., Mialon, M. M., Lardy, R., Silberberg, M., Koko, J. and Veissier, I. 2020. Machine learning to detect behavioural anomalies in dairy cows under subacute ruminal acidosis. *Computers and Electronics in Agriculture* 170, 105233.

Wang, J., Zhang, H., Ji, J., Zhao, K. and Liu, G. 2019. Development of a wireless measurement system for classifying cow behavior using accelerometer data and location data. *Applied Engineering in Agriculture* 35(2), 135–147.

Webster, J. 2008. *Animal Welfare: Limping towards Eden*. Wiley, U.K.

Werner, J., Umstatter, C., Leso, L., Kennedy, E., Geoghegan, A., Shalloo, L., Schick, M. and O'Brien, B. 2019. Evaluation and application potential of an accelerometer-based collar device for measuring grazing behavior of dairy cows. *Animal: An International Journal of Animal Bioscience* 13(9), 2070–2079.

Williams, L. R., Bishop-Hurley, G. J., Anderson, A. E. and Swain, D. L. 2019. Application of accelerometers to record drinking behaviour of beef cattle. *Animal Production Science* 59(1), 122–132.

Winckler, C. 2019. Assessing animal welfare at the farm level: do we care sufficiently about the individual? *Animal Welfare* 28(1), 77–82.

Zambelis, A., Wolfe, T. and Vasseur, E. 2019. Technical note: validation of an ear-tag accelerometer to identify feeding and activity behaviors of tiestall-housed dairy cattle. *Journal of Dairy Science* 102(5), 4536–4540.

Chapter 6

Developing effective training and certification schemes for improving on-farm dairy cattle welfare

Antoni Dalmau and Antonio Velarde, IRTA, Spain

1 Introduction

Animal welfare can be defined in a number of different ways, but there is a growing consensus that whatever the definition, it has to include three elements: the emotional state of the animal, its biological functioning and its ability to show normal patterns of behaviour (Manteca et al., 2009). The five freedoms developed by the Farm Animal Welfare Council (FAWC, 1992) combine elements from the three approaches to welfare explained above and are a very useful framework to identify the main welfare problems as well as a starting point to identify the main welfare components. These freedoms, which represent ideal states rather than actual standards for animal welfare, include: freedom from hunger and thirst, freedom from discomfort, freedom from pain, injury and disease, freedom to express normal behaviour, and freedom from fear and distress.

Monitoring systems, certification schemes and legislation largely rely on the examination of inputs, 'what' or 'how much' of different resources are given to animals (i.e. transport and lairage design, stunning equipment, space requirements, etc.). However, these measures have often been criticized for potentially low validity due to their indirect nature and complex interactions with other resource and management conditions (Waiblinger et al., 2001). Thus, input measures are a poor guarantee for good animal welfare, as animals may

http://dx.doi.org/10.19103/AS.2020.0084.07

experience the same situation or handling procedure differently depending on their genetic background, temperament or previous experiences. In fact, a distinction between animal protection (what people are allowed to do to animals) and animal welfare (probing the animal's own experience of its situation) has grown, and it is now accepted that animal welfare science is largely about the assessment of the animal's own experience (Miele et al., 2013).

A good example is the Welfare Quality Assessment (Botreau et al., 2007) based on the fact that since welfare is a condition of the individual animal, wherever possible, it emphasizes animal-based measures (also called 'outcome' or 'performance' measures) rather than the resource and management in an attempt to estimate the actual welfare state of the animals. Such physiological, health and behavioural measures have inherent advantages over input measures. The first advantage is clearly that, since welfare is a condition of the animal, outcome measures are likely to be the most direct reflection of their actual welfare state. It permits to evaluate the welfare by directly observing the animal, regardless of how and where it is living. Secondly, it is applicable to any production system, as, for instance, an animal with lameness will show the same signs in one country or production system than in another, and remain more transparent to stakeholders. Animal-based measures are evaluative, obtained in a precise way and usually quantitative.

On the other hand, since the start of the modern animal welfare debate (post-1960s) the focus has been mainly on negative aspects of welfare, but there is increasing scientific acceptance of animals having the capacity for experiencing positive emotions (Lawrence et al., 2018). Scientific interest in 'positive welfare' has gained momentum with writings such as Bracke and Hopster (2006) and Yeates and Main (2008). In fact, the same year that Welfare Quality was presenting their final results (the welfare assessment protocols for cattle, pigs and broilers/laying hens), the FAWC from the UK wrote an influential report on the future of farm animal welfare (FAWC, 2009) in which the five freedoms were substituted by the concept of life worth-living. In both cases welfare quality and FAWC, the animal-based measures, are essential.

2 Selection of measures

The first point before creating a certification scheme is to be sure of which is the final objective. Is it the final objective to let the animals decide, or is it the final objective to let the humans decide? In the first case, the assessment is centred on the animals and in the second is centred in what affects animals: environment and management. A vast majority of the certification schemes that exist nowadays in the world in the field of animal welfare are based on management and resources. Only a few of them are beginning to include some animal-based measures, and in most of the cases it represents only a small part of the total measures used.

What is happening with animal welfare certification schemes could be better understood with an example: Imagine that a consumer is going one day to the supermarket and finds a stamp on the door explaining that the supermarket is certified on customer satisfaction. The consumer asks the manager on what this certification is based and the manager explains that there are 10 possible points in total: two points depend on the temperature (not too cold, not too hot), two points depend on the corridor width (not too narrow), two points depend on the cleanliness of floors and walls, two points on the presence of windows with natural light and two points on the presence of clean toilets. The manager finishes explaining that they have 10 points, so they are an excellent supermarket for their customers. The consumer then says goodbye and 'I will never return here'. The manager does not understand anything. He has several papers explaining how important natural light or the cleanliness of floors is for customers. In the meantime, the customer is going back home annoyed by the loud music, poor customer service or how the products are displayed at the store. 'Why don't they ask the customers?', they will suggest. In this case, we easily understand that humans should be the central part of this assessment, but for other animals, welfare certification schemes are usually defined according to what humans are providing to animals instead of how these animals live their lives.

Actually, there are a lot of risk factors that have the potential for improving animal welfare, but a certification system based on these risk factors is avoiding the most important factor of all, the animal that is living there. Although a factor like, for instance, stocking density is for sure a critical issue for animal welfare, the observer is not assessing animal welfare when checking densities. In this case, what is assessed is human decisions, not animal states. For sure, human decisions are easier to assess and to communicate to society. For instance: 'our system does not allow to tie the animals, dehorning cows and ask for 365 days a year with a minimum of 18 hours a day of access to pasture'. Of course, there is a potential here for improving welfare, but the real state of the animals is not assessed, so the impact of these measures (or others) in the real population is just based on the research carried out in another population of animals and, in the worst cases, even just speculations. For this reason, these human decisions should be validated by animal-based measures (Table 1).

If we decided to use animal-based measures, there is a second and more complex problem. Which measures should be selected? Animal-based measures can be collected on-farm either by observation or inspection of the animal or by assessing the effects of a response on the environment, for example, loose faeces on the floor is evidence of diarrhoea in the group. In relation to that, it is important to consider three aspects: validity, repeatability

Table 1 Advantages and disadvantages of using animal-based or resource- and management-based measures for assessing animal welfare

Animal-based	Resource- and management-based
Focussed on the animal state (outputs)	Focussed on what is provided to the animal (input) whether or not the animals use these resources
The different dimensions of animal welfare can be assessed independently	It assesses the availability of the resources but does not allow to assess independently the different dimensions (criteria) of animal welfare
Behaviour and emotions are assessed	Enriched environments, training of farmers and other risk factors to try to avoid pain, fear, distress and abnormal behaviour are assessed
Health and performance are assessed	Risk factors that could induce diseases and injuries are considered
Needs the development of thresholds (i.e. what is acceptable for lameness, dirtiness, mortality, wounds on the body, fear, positive or agonistic behaviour, body condition, etc.)	Can be dichotomic (i.e. tied or not tied animals), with different levels (i.e. disbudding, dehorning, pooled) or in few cases with thresholds (i.e. NH_3 concentration)
More robust and applicable to many context and physical environments (e.g. Nordic and Mediterranean countries)	The assessment should be adapted to the physical environments (e.g. use of straw as bedding material; or need for showers can differ in Finland and Israel)
Not associated to the production system (i.e. an animal is lame in the same way in intensive of organic production systems)	Linked to the production system (i.e. different parameters are used in intensive or organic systems)
Can be used for testing and evaluating new housing and husbandry systems.	The protocol should be adapted to the housing and husbandry systems, and the comparison among them is very limited
Allows to carry out a risk assessment, identification of the main hazards and assesses the magnitude of the impact on the different dimensions of animal welfare.	Identify only the hazard but not the impact on the animal welfare
Requires a relative long farm visit around 6–8 h	Requires a relative short farm visit around 2 h
A proper training is mandatory	A proper training is important
Better in terms of what animal welfare is	Better because of easier, faster and cheaper

and feasibility. Validity is defined as the extent to which we can actually measure what we are supposed to, or in other words, the extent to which a measure is meaningful in terms of providing information on the welfare of an animal or a group of animals. In fact, although welfare is a characteristic of the individual animal, often a group-level animal-based measure is used. For instance, gait scoring is a well-studied animal-based measure of the pain felt by an individual animal when walking, and a group-level measure that is derived from this gait scoring is the percentage of animals in a group that are currently above a certain level of lameness. Validity is sometimes referred to as accuracy, which in a welfare context would be similar to the overall correctness of a measure to identify a specific welfare outcome. For instance, there is a general consensus that vocalizations can be used in several species, including cattle, as an indicator of fearfulness. However, are all the vocalizations of a cow reflecting a state of fearfulness? Are there specific intensities that could be considered a sign of fearfulness compared to others? Are there differences in the way or tendency to vocalize between genders, at different ages, in different breeds? Could an animal that is vocalizing be considered free of fear?

When considering the validity of a measure, an important point to take into account is the presence of confounding factors. Imagine two farms, farm A and farm B, in both cases with a problem of diarrhoea in the animals. Then, a certification company is including the presence of diarrhoea in the assessment of animal welfare and is asking the auditor to check it. The assessor decides then to select a sampling of 50 animals, observes them for some minutes and reports if animals are defecating with diarrhoea; for example, there are 2–5 or there are more than 5. Imagine now that in farm A all the animals are extremely confident with humans due to an excellent human-animal relationship, and in farm B the animals show a great fearfulness to humans. In the first case very few animals or even none depending on the time of the observation will defecate and in the second case, most of them will defecate potentially showing a high prevalence of diarrhoea. The audit will conclude that farm B has a problem with diarrhoea and farm A, does not. Actually, both farms have a problem with diarrhoea, but in farm B there is as well a problem in the human-animal relationship. In the Welfare Quality protocols, vocalizations during the handling of animals in the slaughterhouse are assessed, but only when they are preceded by human contact. In addition, diarrhoea is assessed according to the presence of faeces around the anus and animal defecating with signs of diarrhoea is not considered as such to avoid confounding factors. One way for assessing validity is to study the relationship between a specific welfare outcome indicator and an independently performed assessment of the welfare outcome. The assessment of the validity includes the conditions of how the parameter should be taken. For instance, for how long it is needed to observe a group of animals to determine if the time animals spent on lying down is too much or not. In fact, indicators which are valid at one part of the production cycle may not be applicable in other phases.

A second factor to take into account when selecting a measure is to study its repeatability. Repeatability is defined as the similarity of repeated measurements on the same item (e.g. farm or group of animals). Concerning on-farm welfare assessment, repeatability generally has two components: inter-observer and test-retest repeatability (de Passillé and Rushen, 2005). High inter-observer repeatability is achieved when two (or more) people carrying out an observation of the same animals report similar scores. Test-retest repeatability refers to the chance that the same results will be obtained if the test is repeated. In the special case of welfare assessment, the intra-observer repeatability – which describes the extent to which a single assessor obtains consistent results when evaluating the same animals or farms on different occasions – is part of test-retest repeatability. Test-retest or intra-observer repeatability of measures is usually recorded with a short lapse time between observations, or the same situation is assessed repeatedly from video clips (Martin and Bateson, 2007). If the assessment is not frequently carried out, its consistency over time is especially important when the assessment protocols are going to be used for certification purposes. The results of an assessment tool need to be representative of the long-term situation of a farm detecting significant changes and being not sensitive to minor changes included within the usual variation on farm conditions (Plesch et al., 2010).

The consistency of a welfare assessment should then be studied at different moments of the production cycle when no major changes on the farms occurred. A variety of different methodologies have been used to determine repeatability. In animal welfare science, repeatability of continuous variables is often expressed as a Spearman/Pearson's correlation coefficient (Winckler et al., 2007; Dalmau et al., 2010). In studies where three or more observers measure the same parameter (Wemelsfelder and Millard, 2009; Mullan et al., 2011) or where consistency over time is assessed more than twice (de Rosa et al., 2003), Kendall's coefficient of concordance has been calculated to quantify the overall reliability. The consistency of a test measurement can also be quantified by means of intra-class correlation coefficients (ICC). In some cases, a measure is so difficult to harmonize or to train that is not possible to obtain good repeatability between observers. In these cases, although the parameter could be considered valid, it is not acceptable for a certification scheme. Only with good repeatabilities is it possible to do benchmarking. In this case, the comparison of the same measure between farms with similar housing systems or management practices facilitates the identification of those farms that are outside the normal range of variation in a specific animal-based measure. Such benchmarking of a harmonized set of standardized animal-based measures is especially useful to confirm improvements in welfare following a change.

The third point to take into account is feasibility. In this case it is important to take into account the scenario in which the audit will be performed. For instance, some behavioural studies include techniques such as the open field to study temperament and mood states. The open field would consist in moving the cow to a new environment, usually a closed pen, and observe her behaviour (activity, signs of fear such as defecation, exploratory behaviour, etc.). These methodologies have been used largely in the literature, and so there is some validation and the way to interpret it is well-described, it is possible to obtain good repeatability between observers. However, is it possible to ensure that the conditions of this new environment will be the same for all the cows of the different farms assessed? Will all of them have this new and unexplored facility? Is it possible then to repeat these conditions exactly in all the farms assessed? Actually, this measure would be valid and repeatable, but not feasible.

In addition to the scenario where the assessment will be carried out, other aspects to consider are the methodology used by itself, cost or time. For instance, it is very difficult to assess dehydration in cows by means of external signs. In fact, when it is possible to see this problem in cows, it is because the problem is extremely advanced and probably too late for the animal. However, a way to study a possible risk of dehydration is to assess the haematocrit and presence of albumin in the blood. The question is then if it is feasible to take blood to all the animals in a farm to check haematocrit and albumin in the blood, and, in addition, if the fact of disturbing all the animals with this extraction compensates the result that it will be obtained in a problem that it is supposed to have a low incidence. Finally, sometimes the feasibility is just a question of time and money. Cortisol in hair can be used to assess chronic stress in different species. It is possible to take some hair from cows without disturbing them very much, but is it feasible in terms of cost and time? It will take some days to have the results, and the cost for a sample would stay, at least, at around $15. In this case, of course, what is feasible or not might depend on what the owner of the certification scheme decides to include in the assessment. However, with continued technical developments, especially those associated with automatic recording and precision livestock farming techniques, it is likely that several currently impracticable animal-based measures will become inexpensive and feasible in the future.

Once the feasibility of the measures is individually considered, then it is important to consider the feasibility of all the measures combined. For instance, if it is decided to use 100 measures and once tested in the farm, this takes three full days per farm, this could be considered a feasibility problem. Actually, for the Welfare Quality assessment protocol, the maximum time accepted for an audit on a farm was used for the selection of the final measures for reasons of feasibility. It was decided that an audit should last no longer than 7 h to allow the observers to complete the audit in a day. Then, within these 7 h, and considering all the validated measures with good repeatability, those without overlapping information should be prioritized and those wasting too much time or having other problems of feasibility be abandoned. Practical (including economic) feasibility dictates that the animal welfare assessment system must be as concise, time-effective and easy to carry out as possible (Van Reenen and Engel, 2009). However, to do it properly, it is important to have some guidelines on the dimensions of welfare and how to maintain balance.

3 Dimensions of animal welfare - balance between measures

In determining welfare, there are several domains that are generally regarded as areas of concern and of importance for good animal welfare. A good exercise on how to address these different domains was the Welfare Quality architecture that tried to develop a new way of assessing farm animal welfare, and that was both scientifically rigorous and reflected broader public concerns (Keeling et al., 2013). In this case, The Welfare Quality scientist agreed that in order to assess the overall welfare status of animals, four main areas of concern should be addressed and four related key questions answered. These questions included: are the animals properly fed and watered, are they properly housed, are they healthy and can they show appropriate behaviour? This development evolved into four Welfare Quality principles.

Good feeding includes two elements or criteria: absence of prolonged hunger and absence of prolonged thirst. Hunger may result from malnutrition, undernutrition or both. This causes stress, and, if sufficiently prolonged or severe, this can lead to debilitation, loss of body condition, immunosuppression, disease and an unpleasant emotional state. Thirst is the sensation that accompanies dehydration. Prolonged thirst can occur when animals are given the water of poor quality or when drinking facilities are insufficient or inadequate, and it causes stress, or even debilitation, loss of body condition and disease.

Good housing includes three elements: comfort around resting, thermal comfort and ease of movement. Lack of comfort around resting is likely to reduce resting time. This can lead to at least two major welfare problems. First, the risk of lameness increases if animals receive inadequate rest, and this is particularly important in dairy cattle. Second, animals are often strongly motivated to rest and preventing them from doing so may cause them distress. Farm animals are homeothermic, which means they are able to maintain a relatively constant core body temperature that differs from the environmental temperature within certain limits. Ambient temperatures which are too low or too high cause cold and heat stress, respectively. Cold stress is a particular hazard for newborn animals, and heat stress may result from poor ventilation, inadequate housing and an overly high stocking density. Heat stress is a very common and important welfare problem for dairy cows kept in warm countries (West, 2003). In terms of ease of movement, for instance, the ability of animals to turn around, groom, get up, lie down and stretch their legs has long been considered a basic requisite for good welfare (Brambell Committee, 1965).

Good health is an important component of animal welfare, and it can be defined as the absence of injuries, disease and pain. These negative states can have many causes, including certain management procedures. The legs and feet are the parts of the body that are most frequently injured in farm animals. These injuries interfere with normal behaviour and locomotion and may have a debilitating effect by preventing the animal from feeding normally. A second criterion is the absence of disease. This includes respiratory diseases (such as nasal or ocular discharge or hampered respiration), digestive diseases (bloated rumen, diarrhoea and constipation), reproductive diseases (vulva discharge, metritis) or any other type of disease (neurological, systemic, mastitis, etc.). Finally, another criterion to consider is the use of practices that can cause pain in the animal, such as dehorning or disbudding, tail docking or branding.

The principal appropriate behaviour includes the expression of social behaviour and other important behaviours for the species, good human-animal relationship and a general positive emotional state, that includes the absence of fear. All farm species are social animals and as such are strongly motivated to have contact with conspecifics. Positive social interactions, such as social licking, have a desirable effect on welfare for at least two reasons. First, they have been shown to elicit physiological responses known to be pleasant (Manteca et al., 2009). Second, they reduce the negative effects of stressful events; this is known as 'social buffering' of the stress response (Kikusui et al., 2006). Negative social interactions, such as aggression, impair animal welfare. Aggression may result in injuries, pain and, in extreme cases, the death of the animal. In relation to the second criterion, other important behaviours are based on the fact that animals are strongly motivated to perform particular behaviour patterns. This is the case, for instance, with rooting in pigs, nest building in sows and hens, ground pecking and scratching in poultry, foraging in cattle and exploration in all species. In some circumstances, the inability to perform such behaviour patterns may cause distress and lead to the development of abnormal behaviours, such as stereotypies or redirected behaviours. These could be included in the assessment as indicators of welfare problems. A poor human-animal relationship results in the animals being fearful of the stockperson and other humans. Poor stockmanship is considered the main cause of inadequate human-animal relationships. The term 'stockmanship' covers the way that animals are handled, the quality of their daily management and health care, and how well problems other than disease are recognized and solved (Waiblinger and Spoolder, 2007). Finally, as mentioned previously, the presence of positive emotional states and the absence of negative (such as fear or anxiety) are another key criterion in considering a complete state of welfare. Some animal-based measures may be identified as useful not because they indicate a current welfare problem, but because they are an indicator of potential negative welfare outcomes that are to be avoided (e.g. increased lying time may predict clinical lameness or somatic cell counts can indicate that animal may develop clinical mastitis).

Therefore, the awareness that welfare is multi-dimensional and that its overall assessment requires a multi-criteria evaluation resulted in a scheme that contains the mentioned 12 criteria divided in four principles (Botreau et al., 2007). This was used in the Welfare Quality project to organize the different measures that had been validated, with good repeatability and feasibility. A measure of animal welfare could always be related to at least one of these 12 criteria, so it can be used as a good structure to organize a protocol to ensure that there is a balance in the different dimensions that can affect an animal. Actually, this architecture could be used as well in another way, studying a specific scenario, such as 'transportation of calves in a summer day in a warm country for less than eight hours' and then use the criteria to identify risk factors (i.e. absence of prolonged thirst → high temperatures increasing the needs of water due to evaporation; thermal comfort → due to high temperatures; comfort around resting → deep of bedding material could be reduced in warm conditions, etc.). Finally, an animal-based measure can be used to assess it, such as panting. In risk assessment, it is the negative consequences of the factors that are considered, and so the term hazard is used to mean a factor with the potential to cause poor welfare. Depending on its characteristics (breed, sex, age, etc.), the animal will respond to these inputs and the animal's responses are assessed using animal-based measures. In risk assessment terminology, these responses are the consequences of the factors acting upon the animal. If they are negative they are referred to as adverse effects, and if positive they are referred to as benefits.

4 Training of assessors

Training of assessors is always a key point when a certification scheme is developed. Even in case of management- or resource-based measures, it is important to provide some training and guidelines. For instance, imagine dairy cattle in grassland in a warm country where it is decided that one measure for the welfare of animals will be the presence of shade areas. Apparently, it is easy to assess how much shade is provided to the animals. However, the first question will be if it is only considered the shade provided by natural resources (i.e. trees or bushes) or shade provided by something built by humans. A possible solution is to allow both. The second question is how the parameter will be assessed, just presence of something (i.e. one tree for 200 cows) or a minimum cover of x meters for x% of the herd or even x meters for each animal? The third question is at which time of the day it will be considered, when the sun is in a low position, being the shade the longest, or when the sun is in the highest position, being the shade just equal to the tree's size. The next point will be to consider different kinds of trees to make an estimation when you are far away from a small forest about the real cover this is offering to the animals. In the end, if the objective is to reduce the subjectivity of the measure as much as possible, it is important to train for it and to define it as well, as much as possible, all the possible cases that the assessor will find everywhere.

Another good example is the assessment of water provision. It is easy to define that water must be provided in enough quantity and even asks for a ratio of 10 animals per individual waterer or 6 cm per animal in troughs where more than one animal can drink at the same time. However, what happens when the drinker is shared by two different groups allocated in different pens? Is it the same when animals from both pens can drink at the same time than when the waterer is too narrow and only those from one side can drink? How are the centimetres counted in these cases? You can then consider the cleanliness of the drinkers as another criterion. However, is it the same when the waterer is dirty and the water is clean compared to when both are dirty? It is always really clear if the water is clean or not? Therefore, even the management- and resource-based measures need proper training.

Of course, when animal-based measures are used, the training process is crucial. As mentioned above, the repeatability and feasibility of a measure is essential to be validated in a protocol for a certification scheme. In terms of certification, an animal-based measure should be used only if it is possible to be trained in a reasonable period of time and it can be assessed with good repeatability between observers. There are different reasons for that. The first one is because different people, after a training process, should be able to use the same measure under different scenarios and obtain similar results. The second one is because the same person, under the same scenario and assessing the same animals, should be able to obtain the same results. The third reason is because the farmer who is being audited should be able to understand the measure and if trained, to obtain exactly the same result than the assessor. This last point is especially important and only possible with a robust training programme.

The first point in considering how well an assessor could be trained to take a measure is its degree of difficulty, and this usually will be defined by studying repeatability. For instance, lameness in cattle has been thoroughly investigated. Studies focussed on lameness (or experts in locomotion of cows) use to work with a scale from 1 to 5 points (in some cases with categories such as 1.5, 2.5, 3.5 and 4.5). However, this needs a lot of effort of training and ensuring to see real cases for the different categories in a sufficient number of animals, so it is feasible for a unique measure and a future expert in this matter, but not for an assessor of other 30 measures for certification purposes. In consequence, the parameters used by experts in a topic need to be reduced to less and clearer categories to include the measure in an audit. For instance, in the Welfare

Quality protocol, lameness is assessed by 0, animal walking well; 1, moderate lameness; and 2, severe lameness. This is used for certification purposes in Welfair™ (https://www.animalwelfair.com), bledina (https://www.bledina.com/nos-engagements/) or juustoporti (https://www.juustoportti.fi). For sure, it is not possible to be an expert in cow lameness using just these three categories, but it is enough for a general assessment where it is necessary to reduce the complexity of measures to increase the feasibility of a protocol.

To make a measure simpler to train is not enough in reducing the possible categories, and another important point is to define very clearly the rules. For instance, option 1, a group of animals seems to have a poor body condition or, option 2, this specific animal has three indicators compatible with having a poor body condition. Which one provides clearer rules? Imagine, for instance, that the body of the animal can be divided into different areas to ascertain if the body condition is adequate or not: (1) cavity around tail head; (2) line joining the backbone and hip bones; (3) ends of the transverse processes of the vertebrae; (4) hip bones; (5) tail head; (6) spine; and (7) ribs. Then, you establish a rule such as, assess each area separately and only when you have at least three of these indicators positive (i.e. deep cavity around the tail, end of transverse processes sharp and deep depression between backbone and hip bones) the animal can be considered as too thin. This is a good example of how clear rules could increase repeatability. The reduction of interpretation when assessing animal-based measures will have a high impact on the repeatability of the measure. This is something that must always be highlighted in a training method. You will not be trained in being the best assessors in the world for lameness or body condition, but you will be trained to have the better possible repeatability against yourselves and against other assessors doing the same in other farms for the specific measures trained. If the result of the assessment is an overestimation or underestimation of a problem, this will be addressed by the scientist during the validation and revalidation processes, but not by the individual auditors in the field. If the auditors have the sense that they need to correct something in relation to the rules, the consequence will be a reduction in the repeatability with other assessors, so this must be very clear during the training process: Do not interpret, and just follow the rules.

Once the parameter has been well defined with clear rules, then it is very important that the observer can have access to real cases. To achieve it, a proper training would consist of visiting a farm and using some images or video material as examples. To have good examples with images and videos can be very useful for different reasons. (1) during theoretical explanation, to show some clear examples will help to fix some key concepts; (2) before visiting a farm it is important to do some exercises with a trainer where a set of examples can be seen, discussed and seen again; (3) some extreme, rare or infrequent cases can be trained only having images or videos, as it is difficult to find them in real visits to the farms. In any case, to have a set of images or videos to train an animal-based measure is essential, but it is as well essential to train the measure under real scenarios, for instance, visiting a farm. The visit to see real animals allows us to learn other skills not shown during the classroom course, such as approaching the animals or even in being concentrated enough to remember the rules that every measure needs. Whenever measures are taken from only a sample of all animals in the unit, it is essential that the sample be unbiased and representative (e.g. in terms of sex, age, body size, etc.). It is important to specify how this sample of animals is to be chosen and the number of animals in the sample, and although this can be explained in a room, only in a farm it can be properly understood. It is essential, in consequence, to plan a visit to a real farm if a good training process is required.

Another key point in a proper training is some kind of examination. This process should include a written exam, where the theoretical rules for the different measures, sampling and other issues should be considered. In addition, an exam by means of images or videos, if possible different to those used during training, would also be needed. For this purpose and to avoid deviations from the original definition of a measure it is good to have a set of images assessed by a gold standard (an expert related with the validation process of a measure). Then, each student can be asked to assess the same images and the results compared against the gold standard and only if a minimum agreement is achieved the student is considered well trained. This process should be repeated for all the measures of a protocol. The trainer, in this case, if it is not the expert that validated the measure, can act as a silver standard, helping the student understand the mistake in the images not assessed in the same way that of the gold standard. Finally, a last part of the exam should be a practical exam on a real farm, if possible, different to the farm where the practice was carried out. These three parts should be considered essential for a proper training. Of course, one important point is the number of students per trainer, ideally below 10 and, if possible, never higher than 15, especially for the practice and exam on a farm.

Once trained, it is good to do some extra calibrations. For instance in the Welfair™ certification scheme it is stated that after a training course with the examination by means of a written exam, repeatability with images and practice on farm, a trainer must go with the auditor during the first audit to ensure that no deviations exists (i.e. conduct a shadow audit) and only after this first visit it is considered that the auditor has the ability to apply the protocol properly. Thereafter, the auditor will be recalibrated by a silver standard with a certain frequency and after 5 years a new training and examination process will be needed.

5 Summary and future trends

In order to ensure good welfare in animals, we need both to reduce the risk of poor welfare, by using resource-based and management-based measures, and to monitor actual welfare, by using animal-based measures. The development of certification schemes based on animal-based measures is an on-going process that requires a continuous review and update of the protocol. As not all the criteria are assessed through animal-based measures yet, future research should be focussed on the development and validation of animal-based measures to replace the existing resource and management ones. The research should also include the repeatability, practicability and feasibility of the new measures as well as the calculation of the thresholds and the aggregation in the certification scheme.

Alternative approaches in order to refine the form and duration of the assessment while still ensuring reliable results is also a future challenge. The refinement of a certification scheme should preserve the strength of the integrated approach that lies in the use of the overall assessment protocol, where all the different aspects of animal welfare are covered.

Research might also be on the development of new technologies for the continuous recording of animal welfare measures through sensors, as well as the data processing, analysing and modelling. This information, continuously recorded, has the potential to provide the certification scheme with daily information of the on-farm welfare and reduce the duration of the visits.

In all assessment systems, a critical 'component' is the assessor. Without competent and credible assessors, no certification scheme can function in a way which will satisfy both the stakeholders and the consumers (Velarde and Dalmau, 2012). To achieve high repeatability among assessors, they must be continuously assessed during a robust training course until they develop a uniform scoring. Future research should address the standardization of training courses in scoring the different measures that are to be assessed firstly through classroom presentations and exercises, using photographs and video clips, and then on practical field exercises on farm (Velarde et al., 2011).

Animal welfare science is a fast-growing discipline, and the concept and dimensions might evolve in the future. A certification scheme will need to be updated as new knowledge about welfare dimensions emerge.

6 Where to look for further information

6.1 *Further reading*

- Blokhuis, H., Miele, M., Veissier, I. and Jones, B. (Eds) (2013) *Improving Farm Animal Welfare: Science and Society Working Together: The Welfare Quality Approach*. Wageningen Academic Publishers, Wageningen.

- EFSA Panel on Animal Health and Welfare (AHAW); Statement on the use of animal-based measures to assess the welfare of animals. *EFSA Journal* 2012; 10(6): 2767. [29pp.] doi:10.2903/j.efsa.2012.2767. Available online at: (https://efsa.onlinelibrary.wiley.com/doi/epdf/10.2903/j.efsa.2012.2767).
- Welfare Quality® Assessment protocol for cattle (http://www.welfarequalitynetwork.net/media/1088/cattle_protocol_without_veal_calves.pdf).
- Tucker, C. (Eds) (2007) *Advances in Cattle Welfare*. Woodhead Publishing.
- Martin, P. and Bateson, P. (Eds) (2007) *Measuring Behaviour*. Cambridge University Press, Cambridge.

6.2 Key journals/conferences

- Journal of applied animal behavior science.
- Journal of animal welfare/Journal of animal welfare science.
- Animal/Animals/Journal of Animals Science/ Journal of Dairy Science.
- ISAE annual conference.
- WAFL triannual conference.

6.3 Major international research projects

- www.welfarequalitynetwork.net.
- http://www.fao.org/ag/againfo/themes/animal-welfare/spotlight/es/.
- https://www.dairy.com.au/our-commitments/animal-welfare.
- https://www.oie.int/fileadmin/Home/eng/Health_standards/tahc/current/chapitre_aw_dairy_cattle.pdf.

7 References

Botreau, R., Veissier, I., Butterworth, A., Bracke, M. B. M. and Keeling, L. J. 2007. Definition of criteria for overall assessment of animal welfare. *Animal Welfare* 16(2), 225–228.

Bracke, M. B. M. and Hopster, H. 2006. Assessing the importance of natural behaviour for animal welfare. *Journal of Agricultural and Environmental Ethics* 19(1), 77–89.

Brambell Committee 1965. *Report of the Technical Committee to Enquire Into the Welfare of Animals Kept UnderIntensive Livestock Husbandry Systems*. Her Majesty's Stationery Office, London, UK.

Dalmau, A., Geverink, N. A., Van Nuffel, A., van Steenbergen, L., Van Reenen, K., Hautekiet, V., Vermeulen, K., Velarde, A. and Tuyttens, F. A. M. 2010. Repeatability of lameness, fear and slipping scores to assess animal welfare upon arrival in pig slaughterhouses. *Animal: An International Journal of Animal Bioscience* 4(5), 804–809.

de Passille, A. M. and Rushen, J. 2005. Can we measure human-animal interactions in on-farm animal welfare assessment? Some unresolved issues. *Applied Animal Behaviour Science* 92(3), 193–209.

de Rosa, G., Tripaldi, C., Napolitano, F., Grasso, F., Bisegna, V. and Bordi, V. 2003. Repeatability of some animal related variables in dairy cows and buffaloes. *Animal Welfare* 12(4), 625–629.

Farm Animal Welfare Council (FAWC) 1992. Updates the five freedoms. *Veterinary Record: Journal of the British Veterinary Association* 17, 357.

Farm Animal Welfare Council (FAWC) 2009. *Farm Animal Welfare in Great Britain: Past, Present and Future*. Farm Animal Welfare Council, London, UK.

Keeling, L., Evans, A. Forkman, B. and Kjaernes, U. 2013. Welfare quality: Principles and criteria. In: Blokuis, H., Miere, M., Veissier, I. and Jones, B. (Eds), *Improving Farm Animal Welfare. Science and Society Working Together: The Welfare Quality Approach*. Wageningen Academic Publishers, Wageningen, The Netherlands, pp.91–114.

Kikusui, T., Winslow, J. T. and Mori, Y. 2006. Social buffering: relief from stress and anxiety. *Philosophical Transactions of the Royal Society. B, Biological Sciences* 361(1476), 2215–2228.

Lawrence, A. B., Newberry, R. C. and Spinka, M. 2018. Positive welfare: What does it add to the debate over pig welfare? In: *Advances in Pig Welfare: A Volume in Herd and Flock Welfare*. Elsevier, Oxford, UK.

Manteca, X., Velarde, A. and Jones, B. 2009. Animal welfare components. In: Smulders, F. J. M. and Algers, B. (Eds), *Welfare of Production Animals: Assessment and Management of Risks*. Wageningen Academic Publishers, Wageningen, The Netherlands, pp. 61–77.

Martin, P. and Bateson, P. 2007. *Measuring Behaviour*. Cambridge University Press, Cambridge, UK.

Miele, M., Blokhuis, H., Bennett, R. and Bock, B. 2013. Changes in farming and in stakeholder concern for animal welfare. In: Blokuis, H., Miere, M., Veissier, I. and Jones, B. (Eds), *Improving Farm Animal Welfare. Science and Society Working Together: The Welfare Quality Approach.* Wageningen Academic Publishers, Wageningen, The Netherlands, pp. 19–47.

Mullan, S., Edwards, S. A., Butterworth, A., Whay, H. R. and Main, D. C. J. 2011. Inter-observer reliability testing of pig welfare outcome measures proposed for inclusion within farm assurance schemes. *The Veterinary Journal* 190(2), e100–e109.

Plesch, G., Broerkens, N., Laister, S., Winckler, C. and Knierim, U. 2010. Reliability and feasibility of selected measures concerning resting behaviour for the on-farm welfare assessment in dairy cows. *Applied Animal Behaviour Science* 126(1-2), 19–26.

Van Reenen, K. and Engel, B. 2009. Testing the assessment system: refinement and definition. In: Keeling, L. (Ed.), *An Overview of the Development of the Welfare Quality Project Assessment System*. Cardiff University, Cardiff, UK, pp. 33–42.

Velarde, A. and Dalmau, A. 2012. Animal welfare assessment at slaughter in Europe: moving from inputs to outputs. *Meat Science* 92(3), 244–251.

Velarde, A., Dalmau, A. and Manteca, X. 2011. Innovating on welfare: an added value? In: Bouche, R., Derkimba, A. and Casablanca, F. (Eds), *New Trends for Innovation in the Mediterranean Animal Production*. EAAP129. Wageningen Academic Publishers, Wageningen, The Netherlands, pp. 301–306.

Waiblinger, S., Knierim, U. and Winckler, C. 2001. The development of an epidemiologically based on-farm welfare assessment system for use with dairy cows. *Acta Agriculturae Scandinavica, Section A – Animal Science* 51(suppl. 30), 73–77.

Waiblinger, S. and Spoolder, H. 2007. Quality of stockpersonship. In: Geers, R. and Velarde, A. (Eds), *On Farm Monitoring of Pig Welfare*. Wageningen Academic Publishers, Wageningen, The Netherlands, pp. 159–166.

Wemelsfelder, F. and Millard, F. 2009. Qualitative behaviour assessment. In: Forkman, B. and Keeling, L. (Eds), *Assessment of Animal Welfare Measures for Sows, Piglets and Fattening Pigs*. SLU Service, Wageningen, The Netherlands, pp. 213–219.

West, J. W. 2003. Effects of heat-stress on production in dairy cattle. *Journal of Dairy Science* 86(6), 2131–2144.

Winckler, C., Brinkmann, J. and Glatz, J. 2007. Long-term consistency of selected animal-related welfare parameters in dairy farms. *Animal Welfare* 16(2), 197–199.

Yeates, J. W. and Main, D. C. 2008. Assessment of positive welfare: a review. *The Veterinary Journal* 175(3), 293–300.

Part 3

Improving welfare practices

Chapter 7

Developments in housing of cattle to promote health and welfare

Nigel B. Cook, University of Wisconsin-Madison, USA

1 Introduction

The provision of housing has been an integral component of our management of domesticated cattle for over 150 years. The primary focus of such housing was to provide the animals shelter from climate extremes - principally wet and cold weather during the winter, bedding to keep cattle clean and to provide protection from predation. Housing also served the important purpose of creating a more comfortable environment for the animal caregivers to work in.

The evolution of adult cattle housing in North America has been well reviewed by Bewley, Robertson and Eckelkamp (2017) and the pattern is mirrored in other parts of the world. Tiestall or stanchion barns were used from the mid-1800s to over-winter cattle that would spend the remainder of the year grazing. Over time, this form of housing evolved as a concept with feed mangers and manure gutters that could be used year-round, which is still in use today. Loose-housing 'bedded packs' or 'straw yards' emerged in the 1950s, and while they are still used by many farmers worldwide, they suffer from a requirement for large amounts of bedding material, which makes them costly to operate compared to a form of housing used for dairy cattle that has become the standard since the 1960s, the freestall or 'cubicle' house. Freestall barns

http://dx.doi.org/10.19103/AS.2020.0084.08

for dairy cows were constructed by putting individual stalls into existing loose housing in order to reduce the requirement for bedding, while still keeping the animal clean. Before too long, they were used not only to provide the cows with a place to rest, but also to provide alleys to roam along and access the milking center, and a feed lane along which a ration could be fed. Essentially, the early success of these housing systems allowed cattle to be managed in groups with adaptations to feeding and milking that resulted in a shift from grazing to full-confinement housing for all or part of the year. Bedded packs or open lots remained the mainstay of beef cattle housing, while the increased requirement for hygiene and udder health made tiestalls and freestalls the preferred housing systems for dairy cows. With the continued expansion of herd size, freestalls have emerged as one of the predominant types of dairy cattle housing world-wide, and it will serve as the focus of this chapter (Fig. 1).

The provision of shelter and protection from predation through housing are important components of animal welfare, but for too long the potential negative impacts of housing on basic health and functioning, affective state and the restriction of natural behavior were given low priority by the industry. In our haste to build a barn that was low cost and durable, we forgot that we were still dealing with a grazing animal that lives within a complex social hierarchical

Figure 1 A typical naturally ventilated freestall barn with curtain side walls and deep-bedded sand stalls with fans over the resting area to promote stall use during periods of heat stress.

structure (Beilharz and Zeeb, 1982; Wierenga, 1990). Despite some early research that demonstrated some of the limitations of barn design and group housing on feeding and social behaviors (Kondo and Hurnik, 1990; Miller and Wood-Gush, 1991), the full impact of housing on the cow's behavior, health and productivity was not realized until early this century, with the availability of video recording and activity logger technology, coupled with improved techniques to measure behavioral and health outcomes (see Anderson, 2019).

The last two decades have seen a significant improvement in the way that we house dairy cattle and the aim of this chapter will be to summarize the adaptations and improvements we have made to the freestall housing of adult dairy cattle that have enhanced health and welfare.

2 Natural behaviors

An obvious limitation of housing is that it removes the opportunity for the cow to graze all or part of the year, which is a consistent consumer expectation of dairy herd management (Cardoso et al., 2016). Grazing is clearly an important component of natural living, but it is unclear to what degree housing, which enables other natural behaviors to be displayed, may compensate for an inability to graze. The societal expectation that a modern high-producing dairy cow should graze pasture is likely built on the knowledge of the ancestry of the cow, and the repeated use of images of cows at grass by those who market dairy products that serve to reinforce this expectation. However, studies where cows may choose to remain within a freestall barn or gain access to pasture, show that cows tend to choose pasture during the nighttime, and housing during the daytime (Legrand, von Keyserlingk and Weary, 2009; Charlton et al., 2011). This choice can be impacted by other factors such as weather conditions and milk production (Charlton et al., 2011), with poor weather and higher milk production favoring the choice toward remaining inside the barn. However, cows continue to exhibit this same preference for housing during the day, even when the freestall barn is made less desirable by overstocking (Falk et al., 2012).

It would therefore appear that there are elements of freestall housing that are attractive to the cow, especially during the daytime and during periods of inclement weather that supersede the desire to be outside and graze pasture. These studies also make an important point that not all aspects of natural living are desirable or 'missed' by the cow when the requirement to perform a certain behavior is reduced or removed – such as the desire to graze, or the avoidance of a predator. It is therefore perhaps more salient to focus on the importance of providing the cow the ability to exhibit a broad repertoire of important natural behaviors in our housing systems, rather than emphasizing the need for natural living, which encompasses all types of behavior – and this discussion has recently been made in detail elsewhere (Beaver, Ritter and von Keyserlingk, 2019).

It is common to find housing designs that negatively impact natural behaviors. Dairy cattle naturally exhibit allelomimetic behavior, which is a tendency when one animal in a group exhibits a behavior for the rest of the group to exhibit similar behavior (Stoye, Porter and Stamp Dawkins, 2012). To this day, we continue to construct freestall pens with insufficient feed bunk space for all of the cows to eat simultaneously and to stock them at levels where cattle must compete for a place to rest, even though concerns regarding this approach were voiced almost two decades ago (Grant and Albright, 2001). Similarly, despite the cow being an intensely social animal, the dairy industry constructed facilities that house larger and larger groups of cattle and moved cows between these groups to suit management purposes, even though we have known for some time that regrouping has measurable negative impacts on the cow, particularly those of low social rank (Hasegawa et al., 1997).

While it is true that the dairy cow is very adaptable to circumstances, growing interest in cow behavior and measurable outcomes beyond simply looking at changes in milk production, which is a very poor reflection of welfare, has shown that there is a cost to the well-being of the animal when natural behavior is compromised, particularly with respect to feeding behavior, regrouping, resting behavior, and exposure to climate extremes.

3 Housing adaptations to feeding behavior

Pasture-managed cattle typically graze for around 8–9 h/d (Rind and Phillips, 1999; Tucker et al., 2007; Phillips and Rind, 2001; Beggs et al., 2018), with peaks in activity at dawn and dusk. When cows are confined and fed a total mixed ration (TMR), feeding time is compressed to around 3–5 h/d, spread across 7–12 meals per day (Gomez and Cook, 2010; DeVries, 2019). Feeding activity in freestall-housed systems is more closely related to milking time, fresh feed delivery and feed push-up times (DeVries, von Keyserlingk and Beauchemin, 2003; Mentink and Cook, 2006), putting significant pressure on feed bunk space and usage at these critical times of the day. Indeed, it has been shown that fresh feed delivery in TMR feeding systems is the dominant driver of group feeding activity (DeVries and von Keyserlingk, 2005), prompting debate around the desire to provide sufficient space for all of the cows to eat at the same time, which impacts the type of pen layout used in freestall designs, with either two rows or three or more rows of stalls per feed bunk being the most commonly used.

The case against allowing all of the cows to eat at the same time is essentially an economic one since it is more costly to provide more space per animal. If milk production is the sole outcome measured, then it can be argued that there are many herds with high milk yields that utilize pens with three rows of stalls and less than 0.5 m of bunk space per cow (e.g. Brotzman et al., 2015). However, the case for allowing simultaneous feeding activity has evolved over recent years and is more compelling when we look beyond milk production alone.

First, when cows access a freshly delivered TMR for the first time, they sort against long particles (Leonardi and Armentano, 2003), creating a situation where the first cows to access the feed can consume a different ration than cows that access the feed later on. While this may or may not impact the overall milk production of the group, the difference can certainly impact milk components. For example, one study showed that for every 10% refusal of long forage particles in the ration, milk fat decreased by 0.10% and milk protein decreased by 0.04% points (Miller-Cushon and DeVries, 2017).

Second, competition for space to access feed leads to adaptations to feeding behavior - cows tend to decrease feeding time, increase feeding rate and consume fewer, longer duration, larger meals, with associated less rumination activity, when space is limited (DeVries, 2019). Under such conditions, dominant cows may be unaffected, while subordinate cows show disproportionately altered feeding behavior (DeVries, von Keyserlingk and Weary, 2004) and altered blood metabolites, including elevated non-esterified fatty acids (NEFA) and signs of insulin resistance (Huzzey et al., 2012).

Third, at certain critical times, competitive feeding situations may have a greater negative impact on the cow, and we have argued for improved feed access for housed cows especially during the transition period to optimize health and performance (Cook and Nordlund, 2004; Cook, 2019a). Indeed, cows in a competitive feeding situation prior to calving, that suffered clinical metritis post-partum, were shown to have lower dry matter intake (DMI) during the week before calving compared to healthy cows (Huzzey et al., 2007), and increasing stocking density prior to calving in commercial herds increased the risk for ketosis (Kaufman et al., 2016). However, benefits of increased bunk space may accrue at other times during the lactation cycle; for example, the probability of pregnancy by 150 Days In Milk (DIM) increased linearly as the feeding space per cow increased from 30 cm to 60 cm (Caraviello et al., 2010), and greater bunk space per lactating cow was shown to be associated with greater de-novo fatty acid production measured in bulk tank milk at the herd level (Woolpert et al., 2017).

There is therefore growing evidence to show that there is a detrimental effect of reduced feed space on cow welfare and health. In order for all cows to eat simultaneously, depending on the size of the mature cow and her stage of pregnancy, 24–30 inches (0.61–0.76 m) of feed bunk space per cow are required. Provision of 30 inches (0.76 m) of feed space has become the industry's standard recommendation for the transition period in order to optimize health and productivity (Cook, 2019a; DeVries, 2019), and there is evidence that dairy producers are adopting this recommendation and seeing the rewards (Schnell et al., 2016). This space requirement necessitates the use of freestall pens with two rows of stalls, which may be constructed in a head to head (cows facing each other on a single double-stall platform), tail to tail (two single-stall platforms with one row of cows facing the feed bunk and the other row facing the opposite direction), or head to tail (two single-stall platforms with cows each facing away from the feed bunk) layout, with crossovers between the feed and stall alleys every 15–20 stalls. For breeding pens and other stages of the lactation cycle, 24 inches (0.61 m) of bunk space per cow are recommended to optimize dry matter intake (DMI), also necessitating a two-row pen choice in parlor-milked herds, with crossovers every 20–25 stalls (Cook, 2019b). In robot-milked facilities, the feed bunk may be extended around the robot milking area, increasing feed access and allowing the provision of 24 inches (0.61 m) per cow in a three-row pen design (Pitkäranta et al., 2019).

The design of the feed bunk has also been examined as a way to influence feeding behavior. The use of headlocks at the feed bunk have become commonplace in freestall facilities (Brotzman et al., 2015), used chiefly as a way to facilitate the handling of cattle for their examination and delivery of vaccines and other treatments. However, headlocks also serve to reduce the rate of displacements between cows feeding at the bunk, compared to a simple post and rail design (Huzzey et al., 2006). Their use enables subordinate cows to maintain feeding times, alongside more dominant herd mates (Endres et al., 2005), which may have significance in mixed parity pens, especially during the transition period, where heifers may be housed alongside multiparous cows for the first time. Feed stall dividers have also been employed less frequently at feed bunks to divide and delineate the feed space area and further prevent cows from displacing herd mates, by restricting their access to the shoulder area of the subordinate cow (DeVries and Von Keyserlingk, 2006).

When coupled with adequate feed space, a physical partition at the bunk reduces feed competition and promotes healthy, consistent eating patterns. However, if headlocks or other modifications to the feed space are used, then heifers must be trained to use them first in a low-stress environment during the rearing period.

4 Group management of dairy cattle

As freestall herds have expanded in size, dairy producers began to manage cows in multiple groups, frequently separating breeding or early lactation cows from pregnant or late lactation cows, necessitating a move between these groups once the cow became pregnant or passed peak milk production.

Since cattle are social animals, moving cows between groups creates a stressor on those that are moved as they must re-establish their social rank within the new group. Early experiments focused on regrouping mid- or late- lactation cows and showed elevated frequencies of agonistic interactions between the transferred cow and the other cows in the group for approximately 48 h, with a modest negative impact on milk production of around 3–5% (Brakel and Leis, 1976; Hasegawa et al., 1997). Many dairymen were persuaded to group cattle in a stable pen for the majority of their lactation, which necessitated adaptations to work routines, such as the requirement to breed cows in multiple pens rather than one or a few pens of cows. Others resisted these changes with a primary focus on ease of management. However, it became commonplace to see primiparous cows housed in a separate pen from multiparous cows in larger herds (Brotzman et al., 2015), with the potential for benefits in health and performance as a result (Østergaard, Thomsen and Burow, 2010).

At the turn of this century, we started to become concerned about grouping recommendations for the transition period in larger confinement-housed herds. As cows moved from the dry period to the early lactating milking pens, it was not uncommon for cows to suffer five or more group changes in a period less than two months. We hypothesized that regrouping cows at this critical stage of the lactation cycle, especially during the final weeks of gestation, could carry greater consequences to the health and productivity of the cow (Cook and Nordlund, 2004). Schirmann et al. (2011) were the first to demonstrate the impact of regrouping during the pre-partum period. Moved cows had a 9% decrease in DMI on the day of the move compared to their prior baseline, a finding which supported a recommendation to develop transition housing strategies that decrease the risk of regrouping within the last week prior to calving (Cook and Nordlund, 2004; Cook, 2019a). Studies conducted on socially stable groups of dry cows, compared to groups that received new animals once (Silva et al., 2013, 2014) or twice (Coonen et al., 2011) per week, failed to show measurable improvements in milk production, health, blood metabolites or immune function, but did demonstrate positive effects on feeding behavior in the socially stable group (Lobeck-Luchterhand et al., 2014). Interestingly, it has been subsequently shown that reducing stocking density can mitigate some of the stress of regrouping (Talebi et al., 2014), and it is likely that in the previous studies, the generous feed bunk space provided and other aspects of cow comfort helped to buffer some of the responses expected from regrouping stress (Coonen et al., 2011; Silva et al., 2014).

Over the past decade, a large number of dairy producers have invested in purpose-built transition cow housing and we have observed significant improvements in early lactation health and production as a consequence (Schnell et al., 2016). These producers followed a blueprint to size pens to provide sufficient bunk space and lying space for all of the cows to eat and rest simultaneously, while limiting regrouping during the critical period of 2–7 days before calving. A number of effective transition housing solutions were devised by producers and barn designers to meet these goals, and they have been recently summarized (Cook, 2019a).

These strategies have been coupled with advances in calving pen management that have been spearheaded by understanding the importance of maternity behavior of cows as calving is imminent in housed cattle. As cows managed outdoors enter the first stage of labor, they tend to seek isolation from the rest of the group, presumably as a defense against predators and to promote bonding between the dam and offspring. However, this same type of behavior has been observed in housed cattle and attempts have recently been made to accommodate these behavioral requirements in calving pen design (Proudfoot, 2019). Essentially, we seek to locate the maternity area in a quiet area of the farmstead, free from busy animal and human traffic, close to the dry cow-housing area, and provide cows with the ability to isolate themselves from their herd mates at the point of calving. Calving pens may be designed as individual pens or group pens, with sufficient bedded space being a high design priority (typically 100–150 sq. feet (9.3–13.9 sq. m) per cow). Cows show a preference for straw bedding in the calving area (Campler et al., 2014), but other soft surfaces such as deep sand may suffice when well managed. Isolation may be achieved by providing a wall within the pen to allow cows to rest against the wall or individual areas with solid partitions.

5 Promoting optimal resting behavior

Significant improvements in freestall design have been made over the last two decades, facilitated by the use of video recording to document the precise movements exhibited by the cow when rising and lying down, and capture of lying behavior data on commercial farms and from controlled studies where elements of stall design were manipulated (Tucker, Weary and Fraser, 2003, 2004; Cook and Nordlund, 2005; Anderson, 2019). These improvements have sought to optimize resting behavior and minimize the risk for injury (particularly injury to the hock region of the rear limb) and lameness.

The presence or absence of components of the design of the stall itself can impact lying time. For example, the use of a poorly designed brisket locator can reduce stall use and lying time (Tucker, Zdanowicz and Weary, 2006), and the precise location of the upper and lower rail of the divider loop and neck rail location are important in order to avoid injury and to optimize lying position to promote good hygiene (Cook, 2019b). However, the most significant factors driving stall use documented in the scientific literature are the resting space and the stall resting surface.

Stalls should be appropriately sized to the animal using them, providing sufficient lying area and room to lunge while rising and lying down in order to optimize resting behavior (Tucker, Weary and Fraser, 2004; Solano et al., 2016), and reduce the risk for lameness (Westin et al., 2016). Recommendations for freestall sizing have been made readily available to the dairy industry for over a decade (Cook, 2019b), but the use of existing buildings and compromised stalls continue to create lag in progress with regard to lameness and other injuries (von Keyserlingk et al., 2012).

A variety of freestall surfaces and bedding materials have been employed over time. Traditionally, organic bedding material such as straw or sawdust was used over a simple concrete platform, but these were soon supplemented by the provision of a rubber mat over the concrete. More recently, dairy producers invested in mattresses consisting of thick pads of rubber crumbs, air, water or foam in order to provide increased cushion, while others employed deep loose-bedded stalls, typically utilizing sand as a bedding substrate. Cows bedded on mats and mattresses have been repeatedly shown to be at greater risk for hock injuries compared to deep sand bedding (Barrientos et al., 2013; Chapinal et al., 2014; Cook et al., 2016) and have a higher prevalence of lameness (Cook, 2003; Espejo, Endres and Salfer, 2006; Chapinal et al., 2013; Solano et al., 2015; Cook et al., 2016). The mechanism lessening the injury to the hock with deep loose bedding is easily explained through a reduction in friction and pressure to the skin overlying the bony prominences in the hock region during recumbency. However, the path by which herds with deep loose bedding maintain a lower prevalence of lame cows is more complex. A more comfortable resting surface is likely to promote longer lying times, and this has been documented in some (Gomez and Cook, 2010; Ito et al., 2014; Solano et al., 2016), but not all studies (Tucker, Weary and Fraser, 2003; Norring et al., 2008). Improved cow comfort and adequate time for lying have been thought to play a critical role in claw horn lesion development (Cook and Nordlund, 2009; Proudfoot et al., 2010; Sepúlveda-Varas et al., 2018). However, a more significant effect of deep sand bedding may lie in its ability to optimize resting behavior in cows that become lame, making it easier for cows to rise and lie down and maintain a normal pattern of resting behavior - avoiding abnormally short and abnormally long lying times, which may inhibit recovery and compromise the ability of the cow to continue to milk in the herd (Cook and Nordlund, 2009; Cook, 2019b, 2020). Sand bedding transferred to the concrete floors of the pen also provides for improved traction, reducing the risk for slipping and injury.

Access to a place to lie down is also of great importance in the maintenance of optimal resting behavior. In freestall facilities, access is most commonly impacted by overstocking and in situations when cows spend too long out of the pen milking. While the synchronization of resting behavior is less apparent in housed dairy cattle (Miller and Wood-Gush, 1991), multiple studies confirm that stocking rates in excess of one cow per stall negatively impact lying times (Fregonesi, Tucker and Weary, 2007; Krawczel et al., 2012; Krawczel and Lee, 2019).

Similarly, prolonged time spent milking has a negative effect on the time available for lying down (Gomez and Cook, 2010). Indeed in a study of 111 Canadian freestall-housed dairy herds, workers found that no farm averaged more than 12 h/d lying time when milking time exceeded 3.7 h/d (Charlton et al., 2014). This study also showed a reduction in lying time when milking time exceeded 3.3 h/d and current recommendations are to design facilities, so that time out of the pen milking does not exceed 3 h/d for any cow (Charlton et al., 2014; Cook, 2019b). This is achieved during the design phase of the facility by predicting parlor throughput time and allowing for time to travel to and from the milking center (Cook, 2019b).

6 Responses to extreme climate

While protecting cattle from cold weather is one of the initial driving forces behind the development of housing, the impact of cold extremes on adult cattle behavior and welfare has received little attention. In contrast, the growing concern for the housing of high-producing adult dairy cattle is the management of heat stress.

In hot climates, dairy cows lie down less (Cook et al., 2007; Chen, Schütz and Tucker, 2013, 2016b) and this is believed to influence the seasonal onset of lameness in the late summer (Cook and Nordlund, 2009; Sanders, Shearer and De Vries, 2009). Thermodynamically, cows accumulate heat while resting at a rate of approximately 0.5°C/h and cool at variable rates while standing, depending on environmental circumstances (Hillman, Lee and Willard, 2005; Nordlund et al., 2019). Once core body temperature rises by around 0.4°C, cows are more likely to stand, resulting in shorter-duration lying bouts as THI increases, while the number of lying bouts per day remains relatively constant (Nordlund et al., 2019). Current evidence suggests that this behavioral change occurs despite the implementation of commonly available heat abatement measures such as feed line water soakers (Legrand, Schütz and Tucker, 2011; Chen, Schütz and Tucker, 2013, 2016a) that have been implemented in many progressive freestall-housed dairy herds (Caraviello et al., 2006; Brotzman et al., 2015).

The challenge of improving lying times during periods of heat stress and reducing bunching behavior of cows within barns is an area of active interest. This is fueling a renewed interest in mechanically ventilated facilities employing systems that prioritize the provision of a minimum cooling airspeed of 1-2 m/s in the cows' resting microenvironment (Mondaca, Choi and Cook, 2019) and sufficient exhaust to remove the heat and humidity that accumulate within the

Figure 2 A mechanical cross-ventilated freestall barn with curtain baffles to enhance airspeed in the resting area of the cow. Translucent panels along the end wall enhance the natural lighting within the barn.

barn more effectively than can be done in typical naturally ventilated facilities (Mondaca, 2019; Mondaca, Choi and Cook, 2019). However, because these facilities, configured in tunnel or wide-body cross-ventilated barns (see Fig. 2), tend to be more enclosed, their perception is a challenge for the industry to overcome, and a clear example of where an improvement in housing in one respect can fuel concerns in another.

7 Enhancements to natural behavior

Some recent housing advances are non-structural and serve to enhance the ability of the cow to exhibit natural behaviors. For example, grooming is a natural behavior whereby cows use a substrate such as the bark of a tree to rub against in order to remove dirt and external parasites. Recent work suggests that grooming may just be something cows enjoy, since otherwise clean and healthy cattle were shown to be highly motivated to access a grooming brush (Fig. 3), as much as they were motivated to access fresh feed in a controlled study using a weighted gate (McConnachie et al., 2018). Many dairy producers have installed grooming devices in freestall pens, especially in newly constructed automated

Figure 3 Use of cow brushes has enabled housed cows to engage in natural grooming behavior within a freestall barn.

milking system herds (Salfer, Siewert and Endres, 2018) with apparent positive behavioral effects (DeVries et al., 2007).

8 Conclusion

The story of housing cattle continues to evolve. Positive developments to enhance the health and welfare of the animals have accelerated over the last two decades, driven by the availability of new technology to track behavior and growing interest in its use among the scientific community to answer relevant questions. While consumers of dairy products continue to demand that cows graze pasture, the cow appears to value the benefits of housing in choice studies, particularly when it provides effective shelter from climate extremes and ready access to food and a place to rest.

It is incumbent upon those that design, build and manage housing facilities for dairy cattle to address the health and welfare concerns that such facilities provoke, by embracing current recommendations designed to reduce the risk for injury and lameness, promote health and productivity throughout lactation and enhance the ability of cows to exhibit natural behaviors. The provision of well-designed freestalls, with comfortable deep-bedded stalls, sized to the animals using them, in pens with two-row stall layouts that promote feed access for all of the cows to eat when fresh feed is delivered, within barns designed to protect the cows from extremes of climate, have proven that we can manage housed cattle successfully.

We should not forget that the dairy cow continues to live within a complex hierarchical social system within these buildings, and we need to accommodate these behavioral aspects of a cow's life by limiting regrouping stress and providing alternative technologies such as grooming devices to enhance her ability to exhibit a broad range of natural behaviors when housed.

9 Where to look for further information

Incremental improvements to the housing of dairy cattle and the enhancement of animal welfare is an area of active research and interest globally. In particular, there is significant interest in environmental enrichment, with the merging of housing and outside access systems, and in ventilation and cooling systems that will keep cows comfortable in a warming global environment. Not covered in this chapter, there is significant interest in improvements to calf housing and the replacement of individual pen rearing with pair and group housing management systems. Detailed up-to-date information on the latest trends and recommendations for housing dairy cattle is available at: https://thedairyland initiative.vetmed.wisc.edu/.

10 References

Anderson, N. G. (2019). Introduction: building from the cow up, *Veterinary Clinics of North America - Food Animal Practice* 35(1). Elsevier Inc, 1–9. doi: 10.1016/j.cvfa.2018.10.001.

Barrientos, A. K., Chapinal, N., Weary, D. M., Galo, E. and von Keyserlingk, M. A. (2013). Herd-level risk factors for hock injuries in freestall-housed dairy cows in the northeastern United States and California, *Journal of Dairy Science* 96(6). Elsevier, 3758–3765. doi: 10.3168/jds.2012-6389.

Beaver, A., Ritter, C. and von Keyserlingk, M. A. G. (2019). The dairy cattle housing dilemma: natural behavior versus animal care, *Veterinary Clinics of North America - Food Animal Practice* 35(1). Elsevier Inc, 11–27. doi: 10.1016/j.cvfa.2018.11.001.

Beggs, D. S., Jongman, E. C., Hemsworth, P. E. and Fisher, A. D. (2018). Implications of prolonged milking time on time budgets and lying behavior of cows in large pasture-based dairy herds, *Journal of Dairy Science* 101(11). American Dairy Science Association, 10391–10397. doi: 10.3168/jds.2018-15049.

Beilharz, R. G. and Zeeb, K. (1982). Social dominance in dairy cattle, *Applied Animal Ethology* 8(1–2), 79–97. doi: 10.1016/0304-3762(82)90134-1.

Bewley, J. M., Robertson, L. M. and Eckelkamp, E. A. (2017). A 100-year review: lactating dairy cattle housing management, *Journal of Dairy Science* 100(12). American Dairy Science Association, 10418–10431. doi: 10.3168/jds.2017-13251.

Brakel, W. J. and Leis, R. A. (1976). Impact of social disorganization on behavior, milk yield, and body weight of dairy cows, *Journal of Dairy Science* 59(4), 716–721. doi: 10.3168/jds.S0022-0302(76)84263-4.
Brotzman, R. L., Döpfer, D., Foy, M. R., Hess, J. P., Nordlund, K. V., Bennett, T. B. and Cook, N. B. (2015). Survey of facility and management characteristics of large, Upper Midwest dairy herds clustered by Dairy Herd Improvement records, *Journal of Dairy Science* 98(11), 8245–8261. doi: 10.3168/jds.2014-9264.
Campler, M., Munksgaard, L., Jensen, M. B., Weary, D. M. and von Keyserlingk, M. A. (2014). Short communication: flooring preferences of dairy cows at calving, *Journal of Dairy Science* 97(2). Elsevier, 892–896. doi: 10.3168/jds.2013-7253.
Caraviello, D. Z., Weigel, K. A., Craven, M., Gianola, D., Cook, N. B., Nordlund, K. V., Fricke, P. M. and Wiltbank, M. C. (2010). Analysis of reproductive performance of lactating cows on large dairy farms using machine learning algorithms, *Journal of Dairy Science* 89(12), 4703–4722. doi: 10.3168/jds.S0022-0302(06)72521-8.
Caraviello, D. Z., Weigel, K. A., Fricke, P. M., Wiltbank, M. C., Florent, M. J., Cook, N. B., Nordlund, K. V., Zwald, N. R. and Rawson, C. L. (2006). Survey of management practices on reproductive performance of dairy cattle on large US commercial farms, *Journal of Dairy Science* 89(12). Elsevier, 4723–4735. doi: 10.3168/jds.S0022-0302(06)72522-X.
Cardoso, C. S., Hötzel, M. J., Weary, D. M., Robbins, J. A. and von Keyserlingk, M. A. G. (2016). Imagining the ideal dairy farm, *Journal of Dairy Science* 99(2). American Dairy Science Association, 1663–1671. doi: 10.3168/jds.2015-9925.
Chapinal, N., Barrientos, A. K., von Keyserlingk, M. A., Galo, E. and Weary, D. M. (2013). Herd-level risk factors for lameness in freestall farms in the northeastern United States and California, *Journal of Dairy Science* 96(1). Elsevier, 318–328. doi: 10.3168/jds.2012-5940.
Chapinal, N., Liang, Y., Weary, D. M., Wang, Y. and von Keyserlingk, M. A. (2014). Risk factors for lameness and hock injuries in Holstein herds in China, *Journal of Dairy Science* 97(7), 4309–4316. doi: 10.3168/jds.2014-8089.
Charlton, G. L., Rutter, S. M., East, M. and Sinclair, L. A. (2011). Preference of dairy cows: indoor cubicle housing with access to a total mixed ration vs. access to pasture, *Applied Animal Behaviour Science* 130(1–2). Elsevier B.V., 1–9. doi: 10.1016/j.applanim.2010.11.018.
Charlton, G. L., Haley, D. B., Rushen, J. and de Passillé, A. M. (2014). Stocking density, milking duration, and lying times of lactating cows on Canadian freestall dairy farms, *Journal of Dairy Science* 97(5). Elsevier, 2694–2700. doi: 10.3168/jds.2013-6923.
Chen, J. M., Schütz, K. E. and Tucker, C. B. (2013). Dairy cows use and prefer feed bunks fitted with sprinklers, *Journal of Dairy Science* 96(8). Elsevier, 5035–5045. doi: 10.3168/jds.2012-6282.
Chen, J. M., Schütz, K. E. and Tucker, C. B. (2016a). Cooling cows efficiently with water spray: behavioral, physiological, and production responses to sprinklers at the feed bunk, *Journal of Dairy Science* 99(6), 4607–4618. doi: 10.3168/jds.2015-10714.
Chen, J. M., Schütz, K. E. and Tucker, C. B. (2016b). Sprinkler flow rate affects dairy cattle preferences, heat load, and insect deterrence behavior, *Applied Animal Behaviour Science* 182. Elsevier B.V., 1–8. doi: 10.1016/j.applanim.2016.05.023.
Cook, N. B. (2003). Prevalence of lameness among dairy cattle in Wisconsin as a function of housing type and stall surface, *Journal of the American Veterinary Medical Association* 223(9), 1324–1328. doi: 10.2460/javma.2003.223.1324.

Cook, N. B., Mentink, R. L., Bennett, T. B. and Burgi, K. (2007). The effect of heat stress and lameness on time budgets of lactating dairy cows, *Journal of Dairy Science* 90(4). Elsevier, 1674–1682. doi: 10.3168/jds.2006-634.

Cook, N. B., Hess, J. P., Foy, M. R., Bennett, T. B. and Brotzman, R. L. (2016). Management characteristics, lameness, and body injuries of dairy cattle housed in high-performance dairy herds in Wisconsin, *Journal of Dairy Science* 99(7). Elsevier, 5879-5891. doi: 10.3168/jds.2016-10956.

Cook, N. B. (2019a). Designing facilities for the adult dairy cow during the nonlactation and early lactation period, *Veterinary Clinics of North America - Food Animal Practice* 35(1). Elsevier, 125–138. doi: 10.1016/j.cvfa.2018.10.008.

Cook, N. B. (2019b). Optimizing resting behavior in lactating dairy cows through freestall design, *Veterinary Clinics of North America - Food Animal Practice* 35(1). Elsevier, 93–109. doi: 10.1016/j.cvfa.2018.10.005.

Cook, N. B. (2020). Symposium review: the impact of management and facilities on cow culling rates, *Journal of Dairy Science* 103(4). American Dairy Science Association, 3846–3855. doi: 10.3168/jds.2019-17140.

Cook, N. B. and Nordlund, K. V. (2004). Behavioral needs of the transition cow and considerations for special needs facility design, *Veterinary Clinics of North America - Food Animal Practice* 20(3 SPEC. ISS.), 495–520. doi: 10.1016/j.cvfa.2004.06.011.

Cook, N. B. and Nordlund, K. V. (2005). An update on dairy cow freestall design, *Bovine Practitioner* 39, 29–36.

Cook, N. B. and Nordlund, K. V. (2009). The influence of the environment on dairy cow behavior, claw health and herd lameness dynamics, *Veterinary Journal* 179(3), 360-369. doi: 10.1016/j.tvjl.2007.09.016.

Coonen, J. M., Maroney, M. J., Crump, P. M. and Grummer, R. R. (2011). Short communication: effect of a stable pen management strategy for precalving cows on dry matter intake, plasma nonesterified fatty acid levels, and milk production, *Journal of Dairy Science* 94(5). Elsevier, 2413–2417. doi: 10.3168/jds.2010-3729.

DeVries, T. J. (2019). Feeding behavior, feed space, and bunk design and management for adult dairy cattle, *Veterinary Clinics of North America - Food Animal Practice* 35(1). Elsevier Inc, 61–76. doi: 10.1016/j.cvfa.2018.10.003.

DeVries, T. J., Vankova, M., Veira, D. M. and von Keyserlingk, M. A. (2007). Short communication: usage of mechanical brushes by lactating dairy cows, *Journal of Dairy Science* 90(5). Elsevier, 2241–2245. doi: 10.3168/jds.2006-648.

DeVries, T. J. and von Keyserlingk, M. A. G. (2005). Time of feed delivery affects the feeding and lying patterns of dairy cows, *Journal of Dairy Science* 88(2), 625–631. doi: 10.3168/jds.S0022-0302(05)72726-0.

DeVries, T. J. and Von Keyserlingk, M. A. G. (2006). Feed stalls affect the social and feeding behavior of lactating dairy cows, *Journal of Dairy Science* 89(9). Elsevier, 3522–3531. doi: 10.3168/jds.S0022-0302(06)72392-X.

DeVries, T. J., von Keyserlingk, M. A. G. and Beauchemin, K. A. (2003). Short communication: diurnal feeding pattern of lactating dairy cows, *Journal of Dairy Science* 86(12), 4079–4082. doi: 10.3168/jds.S0022-0302(03)74020-X.

DeVries, T. J., von Keyserlingk, M. A. G. and Weary, D. M. (2004). Effect of feeding space on the inter-cow distance, aggression, and feeding behavior of free-stall housed lactating dairy cows, *Journal of Dairy Science* 87(5), 1432–1438. doi: 10.3168/jds.S0022-0302(04)73293-2.

Endres, M. I., DeVries, T. J., von Keyserlingk, M. A. and Weary, D. M. (2005). Short communication: effect of feed barrier design on the behavior of loose-housed lactating dairy cows, *Journal of Dairy Science* 88(7), 2377–2380. doi: 10.3168/jds.S0022-0302(05)72915-5.

Espejo, L. A., Endres, M. I. and Salfer, J. A. (2006). Prevalence of lameness in high-producing Holstein cows housed in freestall barns in Minnesota, *Journal of Dairy Science* 89(8). Elsevier, 3052–3058. doi: 10.3168/jds.S0022-0302(06)72579-6.

Falk, A. C., Weary, D. M., Winckler, C. and von Keyserlingk, M. A. (2012). Preference for pasture versus freestall housing by dairy cattle when stall availability indoors is reduced, *Journal of Dairy Science* 95(11). Elsevier, 6409–6415. doi: 10.3168/jds.2011-5208.

Fregonesi, J. A., Tucker, C. B. and Weary, D. M. (2007). Overstocking reduces lying time in dairy cows, *Journal of Dairy Science* 90(7). Elsevier, 3349–3354. doi: 10.3168/jds.2006-794.

Gomez, A. and Cook, N. B. (2010). Time budgets of lactating dairy cattle in commercial freestall herds, *Journal of Dairy Science* 93(12). Elsevier, 5772–5781. doi: 10.3168/jds.2010-3436.

Grant, R. J. and Albright, J. L. (2001). Effect of animal grouping on feeding behavior and intake of dairy cattle, *Journal of Dairy Science* 84. Elsevier, E156–E163. doi: 10.3168/jds.S0022-0302(01)70210-X.

Hasegawa, N., Nishiwaki, A., Sugawara, K. and Ito, I. (1997). The effects of social exchange between two groups of lactating primiparous heifers on milk production, dominance order, behavior and adrenocortical response, *Applied Animal Behaviour Science* 51(1–2), 15–27. doi: 10.1016/S0168-1591(96)01082-9.

Hillman, P. E., Lee, C. N. and Willard, S. T. (2005). Thermoregulatory responses associated with lying and standing in heat-stressed dairy cows. *Transactions of the American Society of Agricultural Engineers* 48(2), 795–801.

Huzzey, J. M., DeVries, T. J., Valois, P. and von Keyserlingk, M. A. (2006). Stocking density and feed barrier design affect the feeding and social behavior of dairy cattle, *Journal of Dairy Science* 89(1), 126–133. doi: 10.3168/jds.S0022-0302(06)72075-6.

Huzzey, J. M., Nydam, D. V., Grant, R. J. and Overton, T. R. (2012). The effects of overstocking Holstein dairy cattle during the dry period on cortisol secretion and energy metabolism, *Journal of Dairy Science* 95(8). Elsevier, 4421–4433. doi: 10.3168/jds.2011-5037.

Huzzey, J. M., Veira, D. M., Weary, D. M. and von Keyserlingk, M. A. (2007). Prepartum behavior and dry matter intake identify dairy cows at risk for metritis, *Journal of Dairy Science* 90(7), 3220–3233. doi: 10.3168/jds.2006-807.

Ito, K., Chapinal, N., Weary, D. M. and von Keyserlingk, M. A. (2014). Associations between herd-level factors and lying behavior of freestall-housed dairy cows, *Journal of Dairy Science* 97(4), 2081–2089. doi: 10.3168/jds.2013-6861.

Kaufman, E. I., LeBlanc, S. J., McBride, B. W., Duffield, T. F. and DeVries, T. J. (2016). Short communication: association of lying behavior and subclinical ketosis in transition dairy cows, *Journal of Dairy Science* 99(9), 7473–7480. doi: 10.3168/jds.2016-11185.

Kondo, S. and Hurnik, J. F. (1990). Stabilization of social hierarchy in dairy cows, *Applied Animal Behaviour Science* 27(4), 287–297. doi: 10.1016/0168-1591(90)90125-W.

Krawczel, P. D., Klaiber, L. B., Butzler, R. E., Klaiber, L. M., Dann, H. M., Mooney, C. S. and Grant, R. J. (2012). Short-term increases in stocking density affect the lying and social behavior, but not the productivity, of lactating Holstein dairy cows, *Journal of Dairy Science* 95(8). Elsevier, 4298–4308. doi: 10.3168/jds.2011-4687.

Krawczel, P. D. and Lee, A. R. (2019). Lying time and its importance to the dairy cow: impact of stocking density and time budget stresses, *Veterinary Clinics of North America - Food Animal Practice* 35(1), 47–60. doi: 10.1016/j.cvfa.2018.11.002.

Legrand, A. L., von Keyserlingk, M. A. G. and Weary, D. M. (2009). Preference and usage of pasture versus free-stall housing by lactating dairy cattle, *Journal of Dairy Science* 92(8). Elsevier, 3651–3658. doi: 10.3168/jds.2008-1733.

Legrand, A., Schütz, K. E. and Tucker, C. B. (2011). Using water to cool cattle: behavioral and physiological changes associated with voluntary use of cow showers, *Journal of Dairy Science* 94(7). Elsevier, 3376–3386. doi: 10.3168/jds.2010-3901.

Leonardi, C. and Armentano, L. E. (2003). Effect of quantity, quality, and length of alfalfa hay on selective consumption by dairy cows, *Journal of Dairy Science* 86(2). Elsevier, 557–564. doi: 10.3168/jds.S0022-0302(03)73634-0.

Lobeck-Luchterhand, K. M., Silva, P. R., Chebel, R. C. and Endres, M. I. (2014). Effect of prepartum grouping strategy on displacements from the feed bunk and feeding behavior of dairy cows, *Journal of Dairy Science* 97(5). Elsevier, 2800–2807. doi: 10.3168/jds.2013-7401.

McConnachie, E., Smid, A. M. C., Thompson, A. J., Weary, D. M., Gaworski, M. A. and von Keyserlingk, M. A. G. (2018). Cows are highly motivated to access a grooming substrate, *Biology Letters* 14(8), 1–4. doi: 10.1098/rsbl.2018.0303.

Mentink, R. L. and Cook, N. B. (2006). Short communication: feed bunk utilization in dairy cows housed in pens with either two or three rows of free stalls, *Journal of Dairy Science* 89(1), 134–138. doi: 10.3168/jds.S0022-0302(06)72076-8.

Miller, K. and Wood-Gush, D. G. M. (1991). Some effects of housing on the social behaviour of dairy cows, *Animal Science* 53(3), 271–278. doi: 10.1017/S0003356100020262.

Miller-Cushon, E. K. and DeVries, T. J. (2017). Short communication: associations between feed push-up frequency, feeding and lying behavior, and milk yield and composition of dairy cows, *Journal of Dairy Science* 100(3). American Dairy Science Association, 2213–2218. doi: 10.3168/jds.2016-12004.

Mondaca, M. R. (2019). Ventilation systems for adult dairy cattle, *Veterinary Clinics of North America - Food Animal Practice* 35(1). Elsevier Inc, 139–156. doi: 10.1016/j.cvfa.2018.10.006.

Mondaca, M. R., Choi, C. Y. and Cook, N. B. (2019). Understanding microenvironments within tunnel-ventilated dairy cow freestall facilities: examination using computational fluid dynamics and experimental validation, *Biosystems Engineering* 183. Elsevier Ltd, 70–84. doi: 10.1016/j.biosystemseng.2019.04.014.

Nordlund, K. V., Strassburg, P., Bennett, T. B., Oetzel, G. R. and Cook, N. B. (2019). Thermodynamics of standing and lying behavior in lactating dairy cows in freestall and parlor holding pens during conditions of heat stress, *Journal of Dairy Science* 102(7). American Dairy Science Association, 6495–6507. doi: 10.3168/jds.2018-15891.

Norring, M., Manninen, E., de Passillé, A. M., Rushen, J., Munksgaard, L. and Saloniemi, H. (2008). Effects of sand and straw bedding on the lying behavior, cleanliness, and hoof and hock injuries of dairy cows, *Journal of Dairy Science* 91(2), 570–576. doi: 10.3168/jds.2007-0452.

Østergaard, S., Thomsen, P. T. and Burow, E. (2010). Separate housing for one month after calving improves production and health in primiparous cows but not in multiparous cows, *Journal of Dairy Science* 93(8). Elsevier, 3533–3541. doi: 10.3168/jds.2009-2865.

Phillips, C. J. C. and Rind, M. I. (2001). The effects on production and behavior of mixing uniparous and multiparous cows, *Journal of Dairy Science* 84(11). Elsevier, 2424–2429. doi: 10.3168/jds.S0022-0302(01)74692-9.

Pitkäranta, J., Kurkela, V., Huotari, V., Posio, M. and Halbach, C. E. (2019). Designing automated milking dairy facilities to maximize labor efficiency, *Veterinary Clinics of North America - Food Animal Practice* 35(1), 175–193. doi: 10.1016/j.cvfa.2018.10.010.

Proudfoot, K. L. (2019). Maternal behavior and design of the maternity pen, *Veterinary Clinics of North America - Food Animal Practice* 35(1). Elsevier Inc, 111–124. doi: 10.1016/j.cvfa.2018.10.007.

Proudfoot, K. L., Weary, D. M. and von Keyserlingk, M. A. (2010). Behavior during transition differs for cows diagnosed with claw horn lesions in mid lactation, *Journal of Dairy Science* 93(9). Elsevier, 3970–3978. doi: 10.3168/jds.2009-2767.

Rind, M. I. and Phillips, C. J. C. (1999). The effects of group size on the ingestive and social behaviour of grazing dairy cows, *Animal Science* 68(4), 589–596. doi: 10.1017/S135772980005061X.

Salfer, J. A., Siewert, J. M. and Endres, M. I. (2018). Housing, management characteristics, and factors associated with lameness, hock lesion, and hygiene of lactating dairy cattle on Upper Midwest United States dairy farms using automatic milking systems, *Journal of Dairy Science* 101(9). American Dairy Science Association, 8586–8594. doi: 10.3168/jds.2017-13925.

Sanders, A. H., Shearer, J. K. and De Vries, A. (2009). Seasonal incidence of lameness and risk factors associated with thin soles, white line disease, ulcers, and sole punctures in dairy cattle, *Journal of Dairy Science* 92(7), 3165–3174. doi: 10.3168/jds.2008-1799.

Schirmann, K., Chapinal, N., Weary, D. M., Heuwieser, W. and von Keyserlingk, M. A. (2011). Short-term effects of regrouping on behavior of prepartum dairy cows, *Journal of Dairy Science* 94(5). Elsevier, 2312–2319. doi: 10.3168/jds.2010-3639.

Schnell, B. P., Simons, J. C., Cook, N. B. and Ollivett, T. L. (2016). Transition management practices on Wisconsin freestall dairies, In: Proceedings of *49th Annual Conference of the American Association of Bovine Practitioners* 14(17), Charlotte, NC, 138–139.

Sepúlveda-Varas, P., Lomb, J., von Keyserlingk, M. A. G., Held, R., Bustamante, H. and Tadich, N. (2018). Claw horn lesions in mid-lactation primiparous dairy cows under pasture-based systems: association with behavioral and metabolic changes around calving, *Journal of Dairy Science* 101(10). American Dairy Science Association, 9439–9450. doi: 10.3168/jds.2018-14674.

Silva, P. R. B., Dresch, A. R., Machado, K. S., Moraes, J. G., Lobeck-Luchterhand, K., Nishimura, T. K., Ferreira, M. A., Endres, M. I. and Chebel, R. C. (2014). Prepartum stocking density: effects on metabolic, health, reproductive, and productive responses, *Journal of Dairy Science* 97(9). Elsevier, 5521–5532. doi: 10.3168/jds.2014-8093.

Silva, P. R. B., Moraes, J. G., Mendonça, L. G., Scanavez, A. A., Nakagawa, G., Ballou, M. A., Walcheck, B., Haines, D., Endres, M. I. and Chebel, R. C. (2013). Effects of weekly regrouping of prepartum dairy cows on innate immune response and antibody concentration, *Journal of Dairy Science* 96(12). Elsevier, 7649–7657. doi: 10.3168/jds.2013-6752.

Solano, L., Barkema, H. W., Pajor, E. A., Mason, S., LeBlanc, S. J., Zaffino Heyerhoff, J. C., Nash, C. G., Haley, D. B., Vasseur, E., Pellerin, D., Rushen, J., de Passillé, A. M. and Orsel, K. (2015). Prevalence of lameness and associated risk factors in Canadian Holstein-Friesian cows housed in freestall barns, *Journal of Dairy Science* 98(10). Elsevier, 6978–6991. doi: 10.3168/jds.2015-9652.

Solano, L., Barkema, H. W., Pajor, E. A., Mason, S., LeBlanc, S. J., Nash, C. G. R., Haley, D. B., Pellerin, D., Rushen, J., de Passillé, A. M., Vasseur, E. and Orsel, K. (2016). Associations between lying behavior and lameness in Canadian Holstein-Friesian cows housed in freestall barns, *Journal of Dairy Science* 99(3). Elsevier, 2086–2101. doi: 10.3168/jds.2015-10336.

Stoye, S., Porter, M. A. and Stamp Dawkins, M. (2012). Synchronized lying in cattle in relation to time of day, *Livestock Science* 149(1–2). Elsevier, 70–73. doi: 10.1016/j.livsci.2012.06.028.

Talebi, A., von Keyserlingk, M. A., Telezhenko, E. and Weary, D. M. (2014). Reduced stocking density mitigates the negative effects of regrouping in dairy cattle, *Journal of Dairy Science* 97(3). Elsevier, 1358–1363. doi: 10.3168/jds.2013-6921.

Tucker, C. B., Dalley, D. E., Burke, J. L. and Clark, D. A. (2007). Milking cows once daily influences behavior and udder firmness at peak and mid lactation, *Journal of Dairy Science* 90(4). Elsevier, 1692–1703. doi: 10.3168/jds.2006-577.

Tucker, C. B., Weary, D. M. and Fraser, D. (2003). Effects of three types of free-stall surfaces on preferences and stall usage by dairy cows, *Journal of Dairy Science* 86(2). Elsevier, 521–529. doi: 10.3168/jds.S0022-0302(03)73630-3.

Tucker, C. B., Weary, D. M. and Fraser, D. (2004). Free-stall dimensions: effects on preference and stall usage, *Journal of Dairy Science* 87(5). Elsevier, 1208–1216. doi: 10.3168/jds.S0022-0302(04)73271-3.

Tucker, C. B., Zdanowicz, G. and Weary, D. M. (2006). Brisket boards reduce freestall use, *Journal of Dairy Science* 89(7). Elsevier, 2603-2607. doi: 10.3168/jds.S0022-0302(06)72337-2.

von Keyserlingk, M. A. G., Barrientos, A., Ito, K., Galo, E. and Weary, D. M. (2012). Benchmarking cow comfort on North American freestall dairies: lameness, leg injuries, lying time, facility design, and management for high-producing Holstein dairy cows, *Journal of Dairy Science* 95(12). Elsevier, 7399–7408. doi: 10.3168/jds.2012-5807.

Westin, R., Vaughan, A., de Passillé, A. M., DeVries, T. J., Pajor, E. A., Pellerin, D., Siegford, J. M., Witaifi, A., Vasseur, E. and Rushen, J. (2016). Cow- and farm-level risk factors for lameness on dairy farms with automated milking systems, *Journal of Dairy Science* 99(5). American Dairy Science Association, 3732-3743. doi: 10.3168/jds.2015-10414.

Wierenga, H. K. (1990). Social dominance in dairy cattle and the influences of housing and management, *Applied Animal Behaviour Science* 27(3), 201–229. doi: 10.1016/0168-1591(90)90057-K.

Woolpert, M. E., Dann, H. M., Cotanch, K. W., Melilli, C., Chase, L. E., Grant, R. J. and Barbano, D. M. (2017). Management practices, physically effective fiber, and ether extract are related to bulk tank milk de novo fatty acid concentration on Holstein dairy farms, *Journal of Dairy Science* 100(6). American Dairy Science Association, 5097-5106. doi: 10.3168/jds.2016-12046.

Chapter 8

Advances in understanding behavioral needs and improving the welfare of calves and heifers

Emily Miller-Cushon, University of Florida, USA; and Jennifer Van Os, University of Wisconsin-Madison, USA

1 Introduction

Management of dairy calves and heifers is critical, having the potential to dramatically influence animal welfare and performance during a sensitive period of life and exert longer-term effects on outcomes into lactation. Despite the acknowledged importance of calf rearing, management of dairy calves and heifers is highly variable across the dairy industry. Much research in recent years has furthered our knowledge of calf behavior, having important implications for approaches to manage youngstock housing to improve welfare.

Historically, calf and heifer housing and management decisions have been driven by economic considerations and the welfare and sustainability issue of calf morbidity and mortality in early rearing. However, there have been recent shifts in perspective on calf rearing, including increasing awareness of the influence that early life experiences exert on development, having longer-term implications for both animal productivity and welfare. As such, calf management decisions in recent years have begun to make changes with both short-term and longer-term outcomes in mind. For example, calf milk allowance has historically been restricted to encourage early weaning onto solid feed; yet the practice of

http://dx.doi.org/10.19103/AS.2020.0084.10

providing more biologically appropriate levels of milk during the first weeks of life is becoming more common, in light of evidence of long-term production and health benefits associated with increasing early milk allotment (as reviewed by Khan et al., 2011a).

In addition to increasing longer-term perspectives on optimal housing, calf management decisions are increasingly made with consideration of behavioral needs in mind, due to concern for aspects of animal subjective experience in addition to basic health. This mirrors a shift across all of animal agriculture, where increasing public and industry concern for animal welfare, and research focused on understanding animal behavior and welfare, has been given more priority.

In dairy calves and heifers, consideration of behavioral needs has encompassed aspects of social interactions and feeding, rest and comfort, and opportunities for other behavioral expression, such as grooming and play. More generally, there is growing understanding of the importance of environmental complexity and opportunities for behavioral expression and the role of these early experiences in improving lifelong welfare and reducing behavioral problems, such as the development of abnormal behaviors. Further, understanding of animal behavior and refinement of management to meet behavioral needs increasingly occurs at an individual level, and research in this area reveals individual variability in behavioral development and preferences that should be accommodated to improve welfare. Overall, developing research in this area provides a means to refine management for dairy calf and heifers to improve short- and long-term welfare.

2 Addressing social needs

Under natural conditions, the dam will seek isolation from the herd before calving (Lidfors et al., 1994) such that the calf's initial social contact includes suckling and maternal grooming (Jensen, 2011). Following the return of the calf and cow to the herd, observations of calves reared on pasture suggested that they rest and interact with other calves within the first weeks of life (Sato et al., 1987) and, in natural conditions, long-lasting bonds develop between a female and her female offspring, as well as other females (Reinhardt and Reinhardt, 1981). In the dairy industry, it is common practice to separate the calf from the cow shortly after birth, and subsequent calf housing practices vary widely in social environments and exert broad and long-term effects on behavioral development.

2.1 Maternal contact

Cow-calf separation shortly after birth is a central practice in the dairy industry; yet it is increasingly under scrutiny due to societal concerns that separation negatively affects animal welfare (Ventura et al., 2013). Research in this area has evaluated welfare consequences associated with the timeframe of separating the calf and cow, and recent research efforts have explored approaches and implications to facilitating prolonged cow-calf contact.

It is well established that calves and cows exhibit acute behavioral responses to separation, including vocalizations and increased movement, and evidence suggests that earlier separation (within 24 h, compared to 4 or more days) may reduce vocalizations and time spent looking out of the pen (as reviewed by Meagher et al., 2019). This suggests that distress upon separation may relate to strength of the established bond, and has been used as a basis for recommending prompt cow-calf separation. However, recent research has also evaluated possible behavioral and performance benefits of prolonged maternal contact (Meagher et al., 2019).

There is evidence of bonding in cow-calf pairs, even when suckling is prevented (Johnsen et al., 2015), suggesting that longer-term contact may have a range of developmental benefits for the calf. Prolonged maternal contact supports social development, such as promoting the appropriate expression of submissive behavior (Buchli et al., 2017) and increased expression of play behavior (Flower and Weary, 2001). However, some of these social benefits are not unique to maternal contact and are also seen in calves housed with peers (as discussed in the following section). Prolonged maternal contact is also generally found to reduce development of abnormal oral behaviors, such as cross-sucking and tongue-rolling, compared to calves housed socially but separated from the dam (Roth et al., 2009; Fröberg et al., 2007), which may indicate specific benefits of suckling.

In light of behavioral benefits and societal concerns regarding cow-calf separation, various systems to facilitate prolonged cow-calf contact are being developed. These include free contact, restricted suckling or partial day contact, and use of a nurse or foster cow (Johnsen et al., 2016). However, prolonged cow-calf contact is not currently considered widely feasible, and it remains most common for dairy calves to be housed apart from the dam from shortly after birth.

2.2 Social housing for calves during the pre-weaning period

Once separated from the dam, typical housing systems for dairy calves during the pre-weaning period vary across farms. Calves may be housed individually, which remains most common in many regions, including Canada (Vasseur et al., 2010), the United States (USDA, 2016), and Europe (Marcé et al., 2010), or with same-age social contact, which can take various forms from housing in pairs to larger group pens of calves. Individual rearing became the standard practice in part due to concerns about calf morbidity and mortality, with social isolation seen as reducing the risk of direct calf-to-calf disease transmission and allowing for ease of individual monitoring before the advent of modern automated technologies. However, poor health and performance sometimes observed in large groups of calves was often confounded by low milk allowance (resulting in insufficient nutrient intake and reduced immune function; reviewed by Kahn et al., 2011a) and poor ventilation in older facilities, along with other management risk factors such as poor sanitation, bedding, and dynamic (as opposed to all-in/all-out) grouping practices. Evidence suggests no clear relationship between social housing and disease, whereas management factors such as hygiene, ventilation, nutrition, and calf monitoring are critical with respect to health (as reviewed by Costa et al., 2016a). Rather, there is now a well-established body of evidence indicating broad benefits of social contact during the critical pre-weaning period.

In the last decade, numerous studies have shown that rearing pre-weaned calves in pairs or small groups has many benefits for both animal welfare and performance (Costa et al., 2016a). Social housing improves social and cognitive development, with calves showing improved behavioral flexibility (Gaillard et al., 2014) and willingness to try new feeds (Whalin et al., 2018) when reared in social housing, compared with those reared individually. This translates into better resilience to weaning stress and improved performance during this period (Chua et al., 2002; de Paula Vieira et al., 2010; Duve et al., 2012), although evidence of longer-term effects on behavior and welfare are somewhat less studied to date. Across studies (particularly when fed higher milk allowances), socially housed calves perform at least as well as individually housed calves in terms of solid feed intake, bodyweight at weaning, and average daily gain. Additionally, housing calves in groups while maintaining at least the same space per animal results in an increase in total space, allowing for the expression of a greater range of natural behaviors, including play (Jensen et al., 2015).

Social contact has clear implications for calf affective state. There is broad evidence that calves prefer social contact and seek it out. For example, a study characterizing behavior of pre-weaned calves housed in adjacent outdoor hutches, with a shared outdoor enclosure, found that calves spent over 80% of the indoor time inside the same hutch (Wormsbecher et al., 2017). Although this study was not designed as a formal preference test, the calves' choice to spend time inside the same 2.4 m^2 hutch, at the expense of a larger individual space allowance, suggests that social proximity was a high priority for them. Calves are motivated for social contact, based on willingness to perform an operant task to gain access to another calf, even when only head contact through metal bars was allowed (Holm et al., 2002), suggesting that providing social contact is accommodating a behavioral need. In recent findings, pair-housed calves were found to respond more positively to ambiguous cues, presented in a spatial go/no-go task, compared to individually housed calves (Bučková et al., 2019), demonstrating a more optimistic judgment bias indicative of improved welfare.

Many studies in this area have compared individual and pair housing, suggesting that benefits of social contact are achieved when calves access only a single social partner. However, the needs of calves within larger and more complex social groups have been less explored. A better understanding of social interactions within larger groups of calves may be a means to evaluate individual welfare or refine housing to better accommodate individual behavioral needs. For example, the amount and complexity of physical space provided to group-housed calves may be important for the development and maintenance of social relationships, such as allowing for avoidance of dominant animals or social withdrawal. Social proximity is widely considered an indicator of social bonding, and varies positively with familiarity of social companions (Færevik et al., 2006). Preference for social proximity or isolation, however, may vary between individuals and depend on various factors, including individual personality (Lecorps et al., 2018).

Changes in social proximity have been observed to coincide with morbidity in dairy calves (as modeled using a respiratory disease challenge; Hixson et al., 2018), although it is not well-understood whether changes in social proximity may be due to motivation for social withdrawal or avoidance of a sick calf by pen-mates. Preference for social contact may be better accommodated in group-housed calves through increasing environmental complexity. For example, physical barriers allow for cows to choose to self-isolate during calving (Proudfoot et al., 2014). Recent evidence suggests that dairy calves may seek isolation while in pain, as expressed by increased use of a physical barrier providing visual separation from pen-mates following disbudding (Gingerich et al., 2020). Variability in social behavior between individuals and over time, and welfare implications of accommodating broader expression of social behaviors in the home pen, remain an area of developing research.

2.3 Social dynamics and transitions in weaned heifers

Following weaning, it is common practice on many farms to move heifers to group-housing. For calves previously reared individually, this movement to a social group is a stressful transition that coincides with weaning, such that the separation of these events is preferable (Weary et al., 2008). Housing management during the pre-weaning period may ease this transition, as there is abundant evidence that early life experience has a longer-term effect on behavioral development of dairy calves. Calves reared socially react less fearfully in novel environments (Jensen et al., 1997), indicating an improved ability to adapt to the post-weaning environment. Dairy heifers experience a range of social and housing transitions as they develop, including social regroupings, movement to a freestall barn (as discussed in Section 5.2 of this chapter), and introduction to the milk parlor. Consideration of their behavioral needs through these periods of adaptation and learning may have important welfare implications.

3 Addressing feeding needs

The dairy industry is unique among intensively managed livestock in separating calves at birth from the dam as standard practice. As such, nutritional management from birth is human-controlled, and there are a wide range of approaches for raising calves during their period of consuming milk (or milk substitute) and for providing solid feed and weaning calves from milk.

3.1 Milk feeding of calves

Research concerning needs of dairy calves surrounding milk-feeding has focused on volume, feeding method, and transition from milk to solid feed. The dairy industry has a history of providing dairy calves with restricted milk allotments (less than half of the volume the calf would consume if suckling the cow, or provided free access to milk or milk replacer; Appleby et al., 2001; Jasper and Weary, 2002), with the aim of reducing costs associated with milk-feeding and encouraging solid feed intake to facilitate early weaning. Restricted milk allowances are typically provided at a limited frequency (2–3 times/d), in contrast to feeding which is distributed across the day (Fig. 1) at frequencies of 7–10 times/day for calves with free access to milk (Miller-Cushon et al., 2013a), which mirrors suckling patterns of calves with the cow (reviewed by de Passillé, 2001). Increasing adoption of higher milk allowance feeding plans has been facilitated by the growing use of automated milk feeders, which also provide milk at flexible times, resulting in meal patterning that, in theory, more closely resembles that of a suckling calf. A little-explored potential downside of automatic milk-feeding systems, however, is the restriction of synchronized feeding when only one feeding station is typically provided to groups of 20 or more calves.

A growing body of evidence provides support for more biologically appropriate milk feeding volumes, due to improved performance (including growth, solid feed intake, rumen development, and health, as reviewed by Khan et al., 2011a) and alleviation of the negative affective state of hunger, as indicated by a variety of behavioral responses. Calves vocalize less frequently when provided more milk (Thomas et al., 2001), spend more time lying down and displace other calves from the feeder less (de Paula Vieira et al., 2008), and have increased locomotor play (Jensen et al., 2015). The provision of high milk allowances is appropriate from birth, a time period when the calf has not begun consuming energy from solid feed yet may ingest large quantities of milk (e.g.

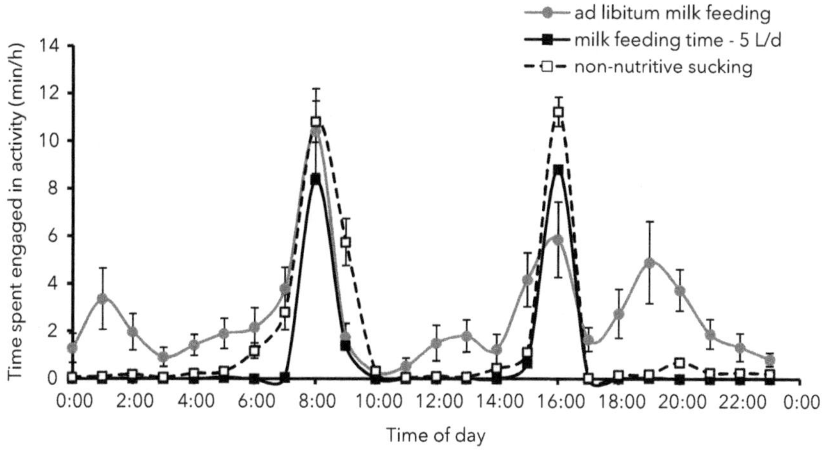

Figure 1 Hourly feeding time for calves provided milk, free access to milk replacer, and calves with restricted access (5 L/d, provided in two meals at 0800 h and 1600 h) and time engaged in teat-directed non-nutritive sucking for calves provided restricted amounts of milk. (Adapted from Miller-Cushon et al., 2013a; reproduced with permission from the Journal of Dairy Science).

averaging 8.6 kg/d in the first week of life, nearly double the allocation of calves provided milk at a conventional restricted rate of 10% of body weight; Appleby et al., 2001).

In addition to milk allowance, feeding methods have consequences for calf behavior and welfare. Obtaining milk through a teat is considered to be rewarding to calves, and has therefore been commonly used as a positive stimulus in studies where calves are placed in testing scenarios (e.g. Horvath et al., 2017; Ede et al., 2018). In practice on some farms, milk allowances are provided in buckets (or trough), requiring the calf to drink or sip the milk, rather than via an artificial teat which accommodates suckling behavior. In most cases, but not all, bucket feeding goes hand in hand with providing restricted milk allowance. It is well established that having an outlet for natural suckling behavior is an important need for calves, with the action of suckling having the function of stimulating hormones involved in digestion and satiety (de Passillé et al., 1993), whereas calves consuming milk from buckets have higher heart rates and increased non-nutritive oral behaviors (Veissier et al., 2002).

In general, performance of non-nutritive oral behaviors may depend on a range of management factors, as discussed throughout this chapter, but are closely related to milk feeding methods and allowance. Cross sucking in particular is considered a nuisance behavior in group-housed calves, with the potential to result in frostbitten ears (in the case of outdoor-housed calves in cold weather) or damaged teats, mastitis, and subsequently reduced milk production (Lidfors and Isberg, 2003); yet not all research indicates that cross-sucking affects udder health (Vaughan et al., 2016). Feeding milk through a teat instead of open buckets can reduce the incidence of cross sucking (de Passillé, 2001; Loberg and Lidfors, 2001; Jensen and Budde, 2006), as can providing a non-nutritive 'dummy' teat to redirect suckling behavior away from conspecifics (de Passillé, 2001; Lidfors and Isberg, 2003), including when calves are fed milk from an open bucket (Veissier et al., 2002). Non-nutritive suckling, including cross sucking, has a periodicity such that the majority of suckling behavior is stimulated when calves taste milk or milk replacer, and this behavior continues after milk is ingested (de Passillé, 2001). Non-nutritive sucking throughout the day in calves provided a restricted milk allowance is shown in Fig. 1, where this behavior occurs for the longest periods around milk meals and surpasses feeding time. To effectively reduce cross sucking, milk-feeding equipment with teats or non-nutritive teats need to remain accessible to the calf for at least 20 min after milk ingestion (Jung and Lidfors, 2001).

In group-housed calves, general social dynamics surrounding milk feeding are important to accommodate. Calves exhibit a high degree of feeding synchrony, even choosing to feed adjacent to a pen-mate where feeding space is limited, despite availability of feeding space elsewhere (Miller-Cushon and DeVries, 2016). This suggests that feeders that thwart social feeding, such as automated milk feeders which provide a stall for one calf, may frustrate feeding motivation. A reduction in feeding synchrony and a high degree of competition is evident even in pair-housed calves provided milk *ad libitum* via a single teat (Miller-Cushon et al., 2014a), yet autofeeders with a single feeding space often accommodate approximately 20 calves on-farm (Jorgensen et al., 2017). Social elicitation of feeding behavior by suckling pen-mates may also explain why cross-sucking has been observed by farmers who provide even high milk allowances to group-housed calves fed by an automated milk feeder, and more research could help elucidate this possibility. Problematic social interactions surrounding feeding, such as displacements, can also be reduced through increasing pen complexity, such as the provision of simple barriers to reduce competition for access to milk (Jensen et al., 2008). However, social feeding appears to be preferred in young calves, and so approaches to facilitate it may be beneficial.

3.2 Weaning from milk

Dairy calves are weaned from milk around 6–8 weeks of age, many months earlier than may occur under more naturalistic settings (e.g. 6 months for beef calves), and the timeframe and method of weaning have important behavioral and performance implications for the dairy calf.

There has been attention placed on refining feeding plans to strike a balance between milk allowance and weaning requirements. Calves provided more milk consume less solid feed prior to weaning, delaying rumen development, and may not maintain consistent growth if weaned too abruptly (reviewed by Khan et al., 2011a). Improvements in growth are seen with gradual reductions of milk allowance (Khan et al., 2007; Sweeney et al., 2010), as well as initiating weaning later (e.g. Eckert et al., 2015; Mirzaei et al., 2018). Cross sucking has been observed to increase directly after weaning, presumably in response to a drop in energy intake (de Passillé et al., 2010), and calves who are better established on solid feed are less likely to cross suck at this time (Keil and Langhans, 2001). Timeframe of weaning impacts both performance and behavior, with calves weaned earlier from a higher milk allowance having more frequent vocalizations and non-nutritive oral behavior (e.g. 6 weeks of age; Eckert et al., 2015).

In recent years, there has been interest in understanding and accommodating the individual needs and behavioral preferences of calves surrounding feeding in the first months of life. Whereas animals are usually managed at a group level, with diets formulated for average nutritional requirements and protocols adopted based on average needs, individuals vary in feeding behavior and preferences. Milk intake and voluntary consumption of solid feed is one area where this variability is particularly apparent. Calves exhibit a wide range of milk intakes and meal patterns when provided free access to milk. Appleby et al. (2001) described *ad libitum* milk intake ranging from 6 to 17 Ld during week 4 of life, with meal frequency ranging from 3 to 22 meals/d. Correspondingly, calves provided milk *ad libitum* also vary in their initiation of solid feed consumption (ranging from 0.01 to 0.31 kg/d during week 4 of life; Appleby et al., 2001). Individual variability in voluntary starter intake during the pre-weaning period has important implications for performance through weaning. Monitoring of individual feeding behavior traits is increasingly facilitated by technology, such as automated milk and starter feeders, and provides opportunity to manage calves at an individual level and accommodate individual differences.

Weaning based on individual starter intake, rather than age, provides a promising approach to individualize management, with potential to reduce cross-sucking during weaning and improve weight gain (Roth et al., 2008). These methods may be particularly useful when calves are provided higher milk allowances. In one study, weaning by reducing a 12 L/d milk allotment as the calf progressively met targets for starter intake (200, 600, 1000, and 1400 g/d; de Passillé and Rushen, 2016) yielded a wide range in duration of weaning (7–49 d) and ages at weaning (58–94 d). Compared to calves weaned by age, this intake-based weaning resulted in reduced milk intake with maintained growth (de Passillé and Rushen, 2016). Benetton et al. (2019) also found that a similar intake-based weaning protocol reduced total milk consumption without negatively impacting growth, compared with weaning based on age. However, those authors also reported that nearly 40% of calves failed to meet some or all of the starter intake targets by the age limit imposed and calves that weaned by intake later (after 63 d) had reduced body weight post-weaning. These findings confirm wide individual variability in early development of feeding behavior and suggest that some calves may still struggle during a weaning transition at this age range, even when their differences in voluntary solid feed intake are somewhat accommodated.

Recent research has further explored predictors of individual variability in behavior outcomes, and evidence suggests that individual variability in feeding behavior through weaning may be explained by differences in personality. Neave et al. (2018) found that increased exploration and activity, as assessed through novel environment, human approach, and novel object tests, was associated with earlier acceptance of solid feed and increased intake through weaning. Evidence of considerable variability in feeding behavior, and associations between feeding behavior and personality traits, indicate that that refining management to meet animal needs necessitates an understanding of individual behavior.

Social environment during the pre-weaning period has implications for acceptance of solid feed and performance through weaning. Social contact during the pre-weaning period stimulates feed intake and reduces feed neophobia (reviewed by Costa et al., 2016a), indicating a function of social contact regarding the development of feeding behavior. Social contact is likely of particular importance for acceptance of solid feed in calves provided higher milk allowances where weaning remains a challenge. In one study, pair-housing stimulated solid feed intake and supported increased growth through weaning in calves provided more milk, whereas pair-housing did not affect performance in calves provided less milk (9 vs. 5 L/d; Jensen et al., 2015). Some evidence suggests that this effect may be enhanced through the provision of more experienced social models. For example, De Paula Vieira et al. (2012) found that housing calves in groups that included an older weaned calf spent more time at the feed bunk and gained more weight prior to weaning. Whereas animals are typically housed by age, especially for pre-weaned calves where vulnerability to disease is a common concern, providing contact with more experienced social models may provide a mechanism to ease dietary and housing transitions that warrant further exploration in weaned heifers.

3.3 Solid feed intake and preferences in young calves

In natural conditions, calves begin to investigate the pasture and consume small amounts in the presence of the dam during the first weeks of life (Nicol and Sharafeldin, 1975). Ruminants have developed the ability to select and learn about a wide range of feed types (Provenza and Balph, 1987), yet under conventional intensive management dairy calves and heifers are typically provided a single mixed diet formulated from a range of ingredients, either pelleted or mixed, potentially alongside some form of forage. Behavioral considerations for calves regarding solid feed provision include palatability and sensory factors, acceptance or selection of feed, variety of feed types provided, and social factors surrounding feeding.

Dairy calves exhibit clear preferences for certain feed ingredients and flavors, which have been studied primarily with the aim of formulating rations that will encourage greater feed intake. For example, in a series of experiments, calves exhibited a consistent preference for soybean meal over other protein sources, increasing their intake of this ingredient in pairwise preference tests and sorting within a mixed diet in favor of that component (Miller-Cushon et al., 2014b). There is no clear evidence to suggest that food palatability influences intake over longer time frames, but there has been some interest in using flavor additives to encourage acceptance of solid feed. Montoro et al. (2011) found that flavoring a starter pellet to match milk replacer during the week prior to weaning had a slight effect in stimulating intake in calves with the lowest pre-weaning intake, suggesting that sensory properties of feed might be influential for calves with lower intakes or, potentially, increased neophobia. In general, the role of sensory properties of feeds in encouraging transition to novel diets has not been well explored.

The variety of feed types and flavors provided to young calves may also be important to consider, to accommodate both individual variability in perceived palatability and possible preference for variety. Cows have more taste receptors on their tongues than humans (Roura et al., 2008), indicating that they are highly sensitive to flavor. In modern dairy production, there is a focus on optimal nutritional formulation of relatively uniform diets, which provide for average nutrient and energy needs, but do not accommodate individual preference or provide opportunity for variety, choice, or sampling and exploratory behaviors. Despite clear evidence of preference for particular flavors and feed ingredients, calves maintain intake of less preferred feed types when offered a choice in experimental settings (Miller-Cushon et al., 2014b). Meagher et al. (2017) also found that weaned heifers selected a fraction of their diet from a rotating offering of a second forage variety or flavored ration, with preference for the varied option subject to individual variability.

This sampling behavior has been attributed to an adaptive function of monitoring the environment (Kyriazakis and Oldham, 1993), but there is also evidence to suggest that ruminants prefer a varied diet. Scott and Provenza (1998) found that lambs preferred a variety of flavors when nutrient content between foods was identical, suggesting that food sensory diversity may have hedonic value. Further, lambs able to choose between feeds that varied over time had lower cortisol levels after feeding than lambs provided a single balanced diet (Catanese et al., 2013). Individual variability in flavor preferences is also well-documented in different species (reviewed by Favreau-Peigné et al., 2013), suggesting that provision of a variety of feed types may better accommodate individual preferences and reduce variability in feed intake. Overall, providing a variety of feed types or flavors may accommodate natural foraging behavior, with potential to improve affective state and stimulate feeding in young calves.

Beyond sensory characteristics of feed, nutrient composition, particle length, and the variety of feed types provided may be important from a behavioral and developmental standpoint. Evidence in young ruminants suggests an ability to associate sensory properties of food with nutrient content, such that provision of a choice of foods (e.g. 'cafeteria' style) may allow individuals to select a balanced diet to meet their needs. For example, lambs appear to select from feed choices to avoid excess protein intake (Kyriazakis and Oldham, 1993). To date, there is limited evidence to suggest that young dairy calves are adept at selecting a balanced solid feed diet from a variety of energy sources to meet specific nutritional demands. For example, calves offered a range of six ingredients were found to over-consume protein (Montoro and Bach, 2012). Similarly, calves also had different protein intakes depending on the protein source, with diet selection apparently driven by differences in perceived palatability rather than nutrient composition (Miller-Cushon et al., 2014c). Whereas dietary selection of solid feeds may not be sensitive to nutrient imbalances in young milk-fed calves, selection of milk relative to high-energy concentrate and forage may be driven by individual preferences and needs. Webb et al. (2014a) described wide individual differences in voluntary intake of bull calves offered free access to milk replacer, concentrate, and various forages, and concluded that diets formulated for average needs may differ from preferences exhibited by most calves. Milk-fed dairy calves do exhibit consistent selection for forage, when offered alongside (shown in Fig. 2) or mixed into starter, and selection of forage may depend on dietary requirements.

Dairy calves are adept at sorting within mixed diets even when particles are small (e.g. diets consisting of different sized pellets, Miller-Cushon et al., 2014c) and will sort mixed diets of forage and starter from the first weeks of life (Miller-Cushon et al., 2013b). Research describing sorting behavior of young calves suggests a motivation to consume forage during the pre-weaning period. For example, calves sorted a mixed diet of 70% texturized grain starter and 30% chopped hay in favor of hay during the pre-weaning period (Miller-Cushon et al., 2013b). Although the individual degree of hay selection has not been associated with rumen environment in young calves, it is well established that provision of hay increases rumen pH and stimulates rumen development, compared to calves provided starter only (Khan et al., 2011b), suggesting that sorting in favor of forage may be motivated by post-ingestive feedback

Figure 2 Image of a pre-weaned calf provided with unchopped hay on a commercial dairy farm in British Columbia, Canada. Photo by Jennifer Van Os.

associated with improved gut health. In adult cattle, it is well established that forage selection (Keunen et al., 2002) and sorting in favor of long ration particles (DeVries et al., 2008, 2014) both increase in response to low rumen pH, suggesting a functional outcome associated with this dietary selectivity. Forage intake in young calves may also be motivated by the sensory process of chewing and ingesting longer feed particles. Dairy calves provided hay have reduced abnormal oral behaviors, such as pen-directed sucking (Fig. 3; Haley et al., 1998; Horvath and Miller-Cushon, 2017), suggesting that consuming hay satisfies a behavioral need for oral manipulation and foraging behavior. Similarly, veal calves provided a diet containing grain and hay performed less tongue rolling compared to calves provided milk only, adjusted to maintain similar weight gain (Webb et al., 2012).

Selection in favor of starter relative to hay increases after weaning off of milk (Miller-Cushon et al., 2013b), suggesting that dairy calves may adjust

Figure 3 Image of a weaned heifer engaging in pen-directed sucking at the University of Florida heifer-raising facility in Alachua, Florida, United States. Photo by Katie Gingerich, University of Florida.

dietary selection to accommodate increasing energy demands when milk is no longer provided. Similarly, weaned bull calves sorted a forage-based total mixed ration in favor of long forage particles and against small ration particles when provided a separate supplemental concentrate, and switched their pattern of sorting in favor of smaller higher energy particles when the supplemental concentrate was no longer provided (Costa et al., 2016b). These results suggest that calves adjust sorting behavior, in favor of particles higher in fiber or energy, depending on their nutrient requirements.

3.4 Feeding behavior in weaned heifers

As they grow, dairy heifers are challenged with many dietary and management changes. Feeding behavior patterns developed early in life have potential to influence longer-term feeding behavior and feed preferences (reviewed by Miller-Cushon and DeVries, 2015), suggesting that feed characteristics during critical pre-weaning and weaning periods may have implications for heifer behavior and welfare, with potential to reduce neophobia and influence

s. For example, in recently weaned heifers, the extent of sorting ed diet in favor of preferred feed components is influenced by the opportunity to perform sorting behavior during the pre-weaning period (Miller-Cushon et al., 2013b). In general, the limited work in this area suggests a range of possible carryover effects of the pre-weaning period on post-weaning feeding behavior, which may affect individual responses to dietary and feed management factors.

As discussed in Section 3.2, social housing during the pre-weaning period also affects development of feeding behavior, enhancing acceptance of novel feeds (Costa et al., 2014), reducing latency to begin feeding once grouped after weaning (De Paula Vieira et al., 2010), and improving competitive success for access to feed (Duve et al., 2012); this therefore has clear implications for the welfare of weaned heifers during key dietary and management transitions. Competition surrounding feeding in the post-weaning environment may also relate to previous experience, as evidence suggest that calves subjected to competitive pressure for access to milk continue to engage in more frequent competitive displacements post-weaning (Miller-Cushon et al., 2014a).

Implications of feeding management for behavior and welfare of weaned heifers have generally received less research attention than calves during the milk-feeding stage. However, an important area of consideration is feeding level and forage provision, as it relates to feeding time and the development of abnormal oral behaviors. For adult cattle, the natural feeding-behavior repertoire includes appetitive components relating to foraging (i.e., searching, selecting) and consummatory aspects to ingesting feed (i.e., chewing, ruminating). In natural settings, cattle graze for a period of time ranging between 7 and 13 h/d (reviewed by Kilgour, 2012). However, intensive housing for cattle typically provides a higher energy, readily accessible diet; consequently, feeding time for confined, *ad libitum*-fed lactating cattle is reported to be between 3 and 5 h (Dado and Allen, 1994; Hosseinkhani et al., 2008).

Although not a standard practice, some work in recent years has evaluated the practice of providing limited amounts of high-energy diets (or 'limit-feeding'), which supports normal growth but decreases feeding time along with manure volume. Limit-fed heifers spent less than 90 min/d feeding in one study (Kitts et al., 2011), and only 28% of the time they would normally spend grazing in another study (Redbo, 1990). High-energy, short-particle diets require less chewing, resulting in less saliva production, which normally helps buffer rumen pH (Beauchemin et al., 2008). Consequently, highly fermentable grain-based diets can result in an excessive level of acid in the rumen, which is not effectively ameliorated by saliva secretion (reviewed by Owens et al., 1998).

Furthermore, feed restriction can affect not only physiological aspects of welfare, but also behavioral outcomes. Tongue rolling in heifers is closely related to feed restriction. For example, Redbo and Nordblad (1997) found that the prevalence of tongue rolling in a group of heifers fed *ad libitum* increased from 33% to 92% when they were fed restricted quantities of a high-energy, low-forage diet. The heifers who exhibited tongue-rolling prior to the period of feed restriction also increased the frequency of tongue-rolling bouts when roughage was restricted. Lindström and Redbo (2000) also determined that, regardless of rumen fill, the ability to forage decreased tongue-rolling, indicating the importance of active foraging behavior for developing heifers. This mirrors effects seen in calves, discussed above, where the provision of forage reduced pen-directed sucking. Behaviour related to feed searching is more common before than after oral stereotypies and tongue-rolling occurs primarily in the hours immediately following a meal (Redbo, 1990), further suggesting a relationship between feeding motivation and tongue-rolling.

The ability to perform the appetitive and consummatory components of feeding behavior may be important for post-weaned cattle. When adult dairy cows' rumens were filled through a fistula (and thus their nutritional needs were met and the sensation of gut fill was achieved, but feeding behavior was limited), they still investigated the empty feed bunk with their noses and tongues as they would during the appetitive component of feeding behavior (Lindström and Redbo, 2000). Cattle may be intrinsically motivated to perform feeding-related behaviors, which are more naturally addressed with high-forage diets or grazing. When fed high-energy, grain-based diets, 7-week-old veal calves, 9-month-old dairy heifers, and 13-month-old beef heifers were motivated to obtain forage, and they performed work by pressing panels and pushing weights, respectively (Webb et al., 2014b; Greter et al., 2015; Van Os et al., 2018). Furthermore, heifers with *ad libitum* access to forage also pushed weights to access additional amounts of the same forage, demonstrating the concept of contrafreeloading (Van Os et al., 2018); this may, in part, reflect an inherent motivation to express foraging behavior, independent from satiation or nutrient requirements.

4 Addressing resting and environmental comfort needs

Lying down is well recognized as an important behavioral need for cattle of all ages. Aspects of the housing environment affecting thermal comfort and the physical comfort of the resting area have important implications for cattle welfare at all life stages.

4.1 Rest and comfort in young calves

Calves spend the majority of their day resting (e.g. 17–18 h/d lying down; Hänninen et al., 2005; Camiloti et al., 2012; Bonk et al., 2013). The type and amount of bedding can influence resting behavior. Calves show clear preferences for dry sawdust over wet sawdust, and an aversion to lying on concrete (Camiloti et al., 2012) and prefer lying on sawdust compared to sand or stones (Worth et al., 2015). Duration of time spent lying has been associated with weight gain (Hänninen et al., 2005), and may depend on bedding comfort. Calves provided wood shavings compared to quarry stones spent more time lying (Sutherland et al., 2013), whereas bedding type in another study did not affect lying time but influenced lying bout frequency (Sutherland et al., 2017).

In addition to increasing the softness of the resting area, bedding can promote thermal comfort, particularly in cooler ambient conditions. Pre-weaned calves, relative to adult cattle, have a higher surface area to volume ratio and generate less heat through ruminal fermentation, and therefore expend energy to stay warm at higher ambient temperature thresholds than older cattle (e.g. 10°C for newborns and 0°C for one-month-old calves; Nordlund and Halbach, 2019). In winter, providing calves with a high plane of nutrition, along with resources such as deep bedding to allow for nesting behavior, can help calves avoid unnecessary energy expenditure to stay warm (Nordlund and Halbach, 2019). Lago et al. (2006) reported an inverse relationship between nesting score, describing the ability of the calf to nestle into the bedding provided, and the prevalence of respiratory disease. In cold weather, calves show lying postures with their legs tucked beneath their bodies in winter (air temperature ≤3°C; Gonzalez-Jimenez and Blaxter, 1962; Brunsvold et al., 1985), likely to reduce heat loss. Calves also prefer to increase proximity to a heat lamp during coldest periods of the day (Borderas et al., 2009).

Conversely, heat stress can also be a concern for young calves. A growing body of research has demonstrated that heat stress experienced in utero can affect dairy calf development, health, and adult productivity (reviewed by Dahl et al., 2019). However, very little research has focused on direct effects of heat stress and heat abatement for calves. While some attempt has been made to characterize temperature-humidity index thresholds for heat stress in calves based on physiological measurements (Kovács et al., 2020) limitations in this area of research to date include a lack of behavioral evaluation or consideration of individual variability.

Limited research to date has explored effects of shade provision and fans as cooling mechanisms for pre-weaned calves. For calves housed in outdoor plastic hutches, although the hutch provides shelter from precipitation, supplemental shade from trees or shade structures is needed to prevent solar gain or a greenhouse effect. Older studies have found benefits of supplemental shade structures over plastic hutches in terms of physiological and growth performance measures (Coleman et al., 1996; Spain and Spiers, 1996). As well, some studies have evaluated the effects of ventilation on physiological and growth responses for calves in outdoor hutches (Moore et al., 2012) or cooled with fans indoors (Hill et al., 2011). More research is warranted on cooling calves housed both indoors and in hutches using fans, natural airflow, water, or the combination, to investigate effects on both behavior and physiology and to evaluate how these interventions interact with effects of heat stress experienced in utero.

4.2 Rest and comfort in weaned heifers

Pregnant heifers are motivated to spend more than half the day lying down (12-13 h/d; Jensen et al., 2005), and when adult cows were presented with a tradeoff between feeding and lying, they prioritized lying (Metz, 1985; Munksgaard et al., 2005). Environmental characteristics have been shown to affect lying time and behavioral responses in cattle of all ages. Because cattle are highly motivated to lie down, they will use surfaces they do not prefer or even find aversive, but will spend less time lying down or show hesitation before doing so (e.g. Krohn and Munksgaard, 1993).

Weaned heifers are commonly housed in open-lot or bedded-pack housing systems (e.g. >60% in the United States; USDA, 2016). Housing and bedding conditions in these systems have important implications for rest and comfort. When pregnant heifers were kept in covered housing on dirt floors with higher moisture content, lying time was reduced by 54% (-7 h) relative to dry conditions (67% vs. 90% soil dry matter; Chen et al., 2017). Similar reductions in lying time were found in studies on muddy open lots or pastures where soil dry matter was not quantified (by 3.6–8.6 h/d in Muller et al., 1996 and by 5.7–6.9 h/d in Fisher et al., 2003). Although lying time increased somewhat on subsequent days of exposure, this was likely out of necessity due to exhaustion, rather than acceptance of the wet lying surface. In addition to the reduction in lying time, heifers avoided lying directly on concrete in dry conditions, but chose to spend the majority of their lying time (≥87%; Chen et al., 2017) on this normally aversive, hard surface in muddy conditions. Together, these findings indicate the importance of providing a dry lying surface to ensure heifers have adequate opportunities to lie down.

As they age and eventually enter the lactating herd, most dairy heifers experience a transition to a freestall barn. For example, in the United States, 15% of weaned dairy heifers are housed in freestalls, which increases to 56% for lactating cows (USDA, 2010). In this system, the designated lying area comprises stalls to orient cattle so they eliminate into an alleyway where manure can be easily cleared away while reducing soiling of the bedding surfaces. This strategy is used to keep both bedding and cattle cleaner than in bedded-pack systems, potentially lowering bedding and labor costs and reducing the percentage of dirty cows (10% vs. 22% in freestalls vs. open lots; USDA, 2010).

Although beneficial from a management standpoint, transition to a freestall barn may be stressful for the heifer. Upon first exposure to sand-bedded stalls after being housed on bedded packs, total lying time was reduced by 20%–29% (3–4 h/d; von Kesyerlingk et al., 2011). In addition, undesirable lying behaviors increased. First, heifers spent more time lying in the alley (O'Connell et al., 1993; Kjæstad and Myren, 2001a; von Keyserlingk et al., 2011), where their bodies directly contact manure, which can increase the risk of developing mastitis (Kjæstad and Simensen, 2001). This pattern across studies is interesting given that heifers often avoid lying down on hard or wet surfaces, such as the alley, as described above. Second, they spent more time perching (i.e. standing with the front feet in the stalls and hind feet in the alley; von Keyserlingk et al., 2011), a behavioral risk factor associated with lameness in adult cattle (Galindo and Broom, 2000; Bernardi et al., 2009). Together, these behaviors suggest that, at least initially, heifers may find stalls to be an unintuitive or unappealing place to lie down. Refusal to lie in stalls may occasionally persist into adulthood: 54% of Norwegian dairy producers, out of 184 surveyed, reporting having a least one cow in the herd who lay in the alley (Kjæstad and Myren, 2001b), although stall design and bedding in the adult housing likely play a role.

To date, only a single study has investigated how to ease the adjustment of heifers to freestalls by covering the concrete stall base with mats or by providing feed in the stalls to attract the heifers into entering them (O'Connell et al., 1993). Although no studies have been published yet, there has been recent interest in mounting small brushes on the stalls to attract the heifers or using experienced social models to demonstrate stall use, similar to how older social companions have been shown to increase feeding frequency and duration (De Paula Vieiera et al., 2012). Some commercial dairies have adopted the strategy of introducing heifers to the components of freestalls gradually, making the transition to a stall base without hardware (i.e. loops, neckrail) before moving heifers to a pen with complete stalls; this stepwise approach merits future study. As well, no studies have investigated the best age at which to introduce cattle to freestalls. On some dairies, the first time they encounter freestalls is after the first calving, which likely presents a detrimental combination of stressors. Conversely, the potentially adverse effects of early life exposure to freestall housing and concrete flooring on behavioral expression as well as physical development merit further investigation.

In terms of thermal comfort, a growing body of evidence has evaluated the effects of heat stress on pregnant heifers and cows during their dry period, with negative implications for their immune function and future fertility and production (Ferreira et al., 2016; Fabris et al., 2019). Studies have found that cooling via water soakers and fans can mitigate these negative fitness and production effects (Ferreira et al., 2016), but less work has evaluated the behavioral effects of cooling on heifers. One study on 10-month-old beef steers naïve to water soakers found that they preferred feed bunks with soakers overhead compared to those without, especially in warmer ambient conditions (Parola et al., 2012). In another warm-weather study of pregnant, non-lactating dairy cattle, some of whom were as young as 16 months old, individual use of a pressure-activated shower varied greatly (Legrand et al., 2011), perhaps in part because the shower was not shaded. Together, these studies suggest that younger cattle with limited or no experience with cooling make some voluntary use of heat-abatement resources, and further study on cooling calves and heifers is warranted.

5 Addressing other behavioral needs

Calves seek stimulation and interaction during their diurnal periods of activity, particularly around feeding, and providing resources to accommodate a range of behaviors, such as grooming and play behavior, may be important for calf welfare.

5.1 Grooming behaviour

Cattle commonly exhibit self-grooming and allogrooming under naturalistic conditions (Reinhardt and Reinhardt, 1981). Aside from licking or scratching themselves with a hind leg, self-grooming can be expressed as rubbing or scratching against objects in the environment (Huber et al., 2008) and it has become increasingly common for dairy producers to provide brushes for adult cows to use for grooming. With motivation testing, lactating cows were shown to put forth great effort to gain access to a rotating mechanical brush (McConnachie et al., 2018), and self-grooming rebounds after a period of restraint (Bolinger et al., 1997). Some rotating mechanical brushes are marketed specifically for youngstock (Fig. 4a) and are used for periods of 20–30 min/d in calves ranging in age from 2 weeks to 7 weeks (Zobel et al., 2017; Horvath and

(a) (b)

Figure 4 Images of (a) pre-weaned calf contacting a rotating brush (DeLaval mini swing brush) at the University of Florida heifer-raising facility in Alachua, Florida, United States (photo by Catherine Hixson, University of Florida) and (b) a post-weaned heifer investigating a non-rotating brush at the University of Wisconsin heifer-raising facility in Marshfield, Wisconsin, United States (photo by Nancy Esser, Marshfield Agricultural Research Station).

Miller-Cushon, 2019). Characterization of brush use suggests that pre-weaned calves visit them for frequent, short periods (e.g. 2-3 min/visit, 10 times/d; Horvath and Miller-Cushon, 2019).

Recently, researchers have also become interested in providing simple, non-rotating brushes (Fig. 4b) to pre- or post-weaned heifers, which may be a practical and cost-effective way to increase opportunities for grooming behavior. Calves may use these manual brushes for less time than rotating brushes (e.g. pre-weaned calves: 4–16 min/12 h observation period; Pempek et al., 2017; 20-week-old weaned heifers: 21–25 min/24 h on average after the first day of exposure, Van Os et al., 2019), but no research to date has directly compared provision of different grooming resources to cattle of any life stage. Further, there has been no investigation of how many of these objects to provide for group-housed cattle or social dynamics surrounding their use.

Provision of non-rotating brushes may be intended to stimulate grooming behavior, but may also serve to more generally increase environmental complexity or serve as an outlet for oral manipulation (Horvath et al., 2020; Van Os et al., 2019), as shown in Fig. 4b. Likewise, when hemp ropes are mounted in the pen, both 2-week-old dairy calves (Zobel et al., 2017) and older feedlot cattle will regularly chew on them. Recent evidence suggests that provision of manual brushes to pre-weaned calves reduces performance of pen-directed abnormal oral behavior and reduces standing time surrounding feeding (Horvath et al., 2020), suggesting that they may decrease arousal during the period of high activity at feeding.

Expression of allogrooming and self-grooming behavior in young calves and heifers may have a range of welfare implications and also serve as a possible indicator of welfare. The calf is groomed by the dam in the first hours after birth (Jensen, 2011) and a reduction in heart rate of receivers of allogrooming (Laister et al., 2011) may suggest that allogrooming has a calming effect. Expression of both allogrooming and self-grooming may provide an indication of calf welfare from both behavioral and physiological perspectives, as these types of grooming both relate to health. Self-grooming was reduced in calves following a bacterial endotoxin challenge (Borderas et al., 2008). Calves challenged with a respiratory disease pathogen initiated less allogrooming but, interestingly, tended to receive more social grooming from healthy penmates (Hixson et al., 2018).

Environmental factors that affect expression of allogrooming in socially housed calves are not well established, as evidence suggests no effect of group size (Færevik et al., 2007) or brush provision (Horvath and Miller-Cushon, 2019) on expression of allogrooming. However, allogrooming is seen more between familiar penmates (Færevik et al., 2007), indicating its role in social bonding and supporting the importance of maintaining stable social groups. Self-grooming is influenced by environmental factors, such as bedding type (e.g. increasing when bedded on rice hulls or sand vs. long wheat straw; Panivivat et al., 2004).

Some evidence suggests that provision of brushes also stimulates self-grooming in pre-weaned calves, in the form of body-directed licking or scratching, and improves coat cleanliness during weaning (Horvath and Miller-Cushon, 2019).

5.2 Play behaviour

Opportunity for play behavior, both locomotor play (e.g. running and jumping) and social play, may serve an important developmental role for dairy calves. Locomotor play behavior may be affected by the amount of physical space provided to pre-weaned calves (Jensen and Kyhn, 2000). However, effects of space allowance and bedding type on running are not consistently reported; Sutherland et al. (2014) described that calves reared with less space spent more time running in a test arena, whereas behavior in the home pen did not differ. Evidence suggests that bedding material affects play behavior, with more locomotor play (running, head shakes, jumps, kicks, and leaps) observed in calves reared on wood shavings compared to stones (Sutherland et al., 2013).

When play behavior is facilitated after a period of restriction, calves exhibit a rebound effect, suggesting that accommodating locomotor play via social contact, physical space, or a combination of both may be considered a behavioral need. For example, calves reared individually spend more time in locomotor play when introduced to a new space and companions after weaning, compared to calves raised in groups (Valníčková et al., 2015). A rebound effect in locomotor play is also seen when calves reared in smaller space allowances are placed in an open-field test after weaning (Jensen and Kyhn, 2000). However, Bertelsen, and Jensen (2019) found that the duration of social play did not increase following a period of deprivation, suggesting that there is not a motivational build-up for all types of play.

Expression of play behavior may reflect the affective state of calves, as it is often found to be associated with negative states such as hunger and pain. Locomotor play behavior decreases when calves are provided restricted amounts of milk (Jensen et al., 2015) and following disbudding (Mintline et al., 2013; Winder et al., 2017). Given the reduction in play behavior coinciding with these negative states, there has been discussion that play may be considered an indicator of positive emotions; however, evidence in this area is lacking to date (as reviewed by Ahloy-Dallaire et al., 2018).

6 Common themes and developing areas of research

6.1 Environmental complexity and animal welfare

We have addressed specific areas of dairy calf and heifer behavioral needs and corresponding management issues and solutions, yet a broader consideration in intensive housing of livestock is opportunity for behavioral variability and environmental complexity. Ultimately, typical on-farm housing for dairy calves and heifers remains quite restrictive, especially in cases of individual housing for pre-weaned calves, but also in the range of behaviors afforded to group-housed calves. Welfare consequences associated with boredom and frustration in restricted environments are well documented, specifically regarding the development of abnormal behaviors (Mason, 1991). Increasing environmental complexity to accommodate greater behavioral variability may have a range of benefits for behavioral development and welfare, and there are gaps in our knowledge regarding resources and optimal housing environments to provide for developing heifers. Here we summarize some areas of ongoing interest in the impact of environmental complexity on abnormal behaviors, cognition, and boredom.

In dairy calves, non-nutritive oral behaviors are prevalent, and relate to feeding management, including milk feeding with a teat (Section 3.1), and provision of forage in young calves (Section 3.3), as well as weaned heifers (Section 3.4). However, non-nutritive sucking can still develop when calves are provided higher milk allowances. Pen-directed sucking and licking behaviors (Fig. 2) occupy a considerable amount of time, for example, 40–60 min/12 h observation periods (Pempek et al., 2017; Horvath et al. 2020), around 10–20 % of the calf's active time (based on lying time of approximately 18 h/d; Bonk et al., 2013).

Provision of items meant to reduce boredom and abnormal behaviors, such as manual brushes and items to chew on, has not consistently been found to have any effect on these behaviors (Pempek et al., 2017). Non-nutritive oral behaviors (e.g. manipulating substrates, tongue-rolling, and cross-sucking) are also common in veal calves and, although veal calves are subject to a range of different management factors compared with replacement dairy heifers, increased space allowance is one factor which reduces the incidence of abnormal oral behaviors (Leruste et al., 2014). Young calves appear to be highly motivated for stimulation during their diurnal periods of activity, and the approach to reduce development of abnormal oral behaviors may depend on substantially increasing environmental complexity and opportunities for behavioral variability.

It is also well established across species that environmental complexity affects cognitive development. In dairy calves, positive effects of social contact on behavioral flexibility, as assessed in a reversal learning task, has been demonstrated repeatedly (Gaillard et al., 2014; Meagher et al., 2015). Evidence also suggests that increasing feeding complexity, through provision of forage and a teat for sucking, may improve reversal learning ability (Horvath et al., 2017; Horvath and Miller-Cushon, 2020). These early opportunities for behavioral variability may have important implications for lifelong welfare. For example, behavioral flexibility during the pre-weaning period has been associated with the ability to adapt to a novel environment after weaning, where calves with greater success during the reversal learning stage of a procedural learning task subsequently had lower latency to begin consuming feed in a novel pen and exhibited more exploratory behavior (Horvath and Miller-Cushon, 2018). It is well established in other species that environmental complexity affects both cognition and development of abnormal repetitive behaviors, with individuals displaying stereotypic behaviors also having reduced cognitive flexibility (reviewed by Lewis et al., 2006).

Although development of abnormal behaviors is of concern and has been relatively frequently measured and considered in research to date, housing animals in restrictive environments has broader implications for animal welfare. As reviewed by Meagher (2019), boredom is an animal welfare concern associated with restrictive and unvarying environments, evidenced by greater generalized motivation for stimuli in animals in more restrictive housing. A relatively barren housing environment may, in part, explain the contrafreeloading performed by 13-month-old beef heifers (Van Os et al., 2018). Performing work to access a resource that is simultaneously freely available may allow animals to express agency (Špinka, 2019), afford them a sense of control over their environment, or relieve boredom. These mechanisms likewise possibly explain why dairy heifers showed greater excitement, as measured by heart rate and locomotor behaviors (bucking, jumping, kicking), when they completed an operant task to obtain feed compared with receiving the reward freely (Hagen and Broom, 2004). Boredom in dairy calves has yet to receive formal study, but poses welfare concerns as an aversive experience that also has negative health consequences, as evaluated in humans and other animal species (Meagher, 2019).

Inactivity in dairy calves has received some attention, and may reflect a response to unvarying or restrictive environments. Webb et al. (2017) found that inactivity was affected by dietary complexity, examined in calves provided diets varying in restriction of feeding level and supplemental forage. These authors observed some indication of a negative association between duration of inactivity, such as standing or lying idly, and abnormal oral behaviors, including tongue play and object manipulation. This relationship has also been observed in other species, and inactivity is suggested as an alternative response to stereotypic behavior for animals in restrictive housing (Fureix et al., 2016). Individual differences in susceptibility to boredom are also apparent in other species (Meagher, 2019) and bear consideration alongside other individually variable responses seen in dairy calves and heifers.

6.2 Accommodating individual differences and providing choice

In most livestock production systems, animals are managed at a group level, with diets formulated based on average nutritional needs, and many management decisions made using age-based protocols, rather than accommodating individual variability. Whereas much research in this area has historically reported group-average behavioral responses, there is growing interest in understanding causes and consequences of individual variability in behavioral traits.

Research in recent years has explored personality traits in dairy calves, a topic which has long been studied in humans but is of increasing interest in animals, and has important implications for individual welfare (reviewed by Richter and Hintze, 2019). In dairy calves, research in this area has revealed associations between personality traits and behavioral responses related to feeding and socialization. In Section 2.2, we briefly discuss individual variability in social behavior, alongside evidence that social proximity to penmates is associated with individual personality traits (Lecorps et al., 2018). Individual variability in social behavior suggests that designing housing to accommodate different preferences has important welfare implications. In Section 3.2 we discuss evidence that personality traits are associated with initiation of solid feed intake (Neave et al., 2018), an individual metric that is widely variable, as well as research exploring weaning dairy calves based on starter intake rather than age. Evidence that starter intake-based weaning protocols reveal a wide age range at time of weaning (Benetton et al., 2019) suggests the importance of accommodating individual variability in implementing this transition.

Accommodating individual differences in expression of certain behaviors may require increased environmental complexity to provide groups of animals with more choices. Beyond accommodating individual differences, animals may specifically benefit from choices in their environment. In Section 3.3, we discuss evidence that young ruminants may prefer varied feed sources (Meagher et al., 2017). Promoting animal agency and providing choices in environmental resources in livestock production settings have been discussed as important means to improve animal welfare and accommodate the needs of individual animals (Špinka, 2019). This relates to our discussion of complexity of social housing in Section 2.2, where we discuss recent findings that calves altered the use of a visual barrier after disbudding. This developing area of research has the potential to improve individual animal experiences, reducing aversive states such as boredom and facilitating a broader range of motivated behaviors, and accommodating individual preferences in ways that may improve growth performance as well as welfare. Further research to apply these concepts to practical housing solutions for dairy calves and heifers will be beneficial.

7 Conclusion

This chapter provides an overview of key concepts related to accommodating behavioral needs of dairy calves and heifers to improve welfare, including social, nutritional, rest and comfort needs, and expression of other behaviors such as grooming and play. Effects of early experience are seen both in short-term investigation of calf development, and in longer-term behavior of heifers after weaning, emphasizing the importance of refining early-life management strategies. Although the focus here has been on behavioral needs and preferences of dairy calves, evidence throughout supports the link between improved growth performance when behavioral needs are met. Areas of developing research and common themes emerging in recent research include generally addressing environmental complexity and understanding and accommodating individual differences. Research towards improving dairy calf and heifer management is a key component of dairy sustainability, amid ongoing societal concern for animal welfare.

8 Where to look for further information

The following textbook chapter provides a useful overview of similar topics related to calf housing and welfare, with a particular focus on health and early life management, including colostrum management:

- Miller-Cushon, E. K., Leslie, K. E. and DeVries, T. J. 2017. Ensuring the health and welfare of dairy calves and heifers. Chapter 6 *in* Achieving sustainable production of milk, Volume 3: Dairy herd management and welfare. Burleigh Dodds Science Publishing, Cambridge, UK.

The following review articles provide good overviews of key subjects discussed within this chapter, including effects of social housing for dairy calves, milk allowance, and environmental enrichment in general:

- Costa, J. H. C., von Keyserlingk, M. A. G. and Weary, D. M. 2016. Invited review: Effects of group housing of dairy calves on behavior, cognition, performance, and health. *J. Dairy Sci.* 99:2453–2467.
- Khan, M. A., Weary, D. M. and von Keyserlingk, M. A. G. 2011. Invited review: Effects of milk ration on solid feed intake, weaning, and performance in dairy heifers. *J. Dairy Sci.* 94:1071–1081.
- Mandel, R., Whay, H. R., Klement, E. and Nicol, C. J. 2016. Invited review: Environmental enrichment of dairy cows and calves in indoor housing. *J. Dairy Sci.* 99:1695–1715.

9 References

Ahloy-Dallaire, J., Espinosa, J. and Mason, G. 2018. Play and optimal welfare: does play indicate the presence of positive affective states? *Behav. Processes* 156:3–15.

Appleby, M. C., Weary, D. M. and Chua, B. 2001. Performance and feeding behaviour of calves on ad libitum milk from artificial teats. *Appl. Anim. Behav. Sci.* 74(3): 191-201.

Beauchemin, K. A., Eriksen, L., Nørgaard, P. and Rode, L. M. 2008. Short communication: salivary secretion during meals in lactating dairy cattle. *J. Dairy Sci.* 91(5):2077-2081.

Benetton, J. B., Neave, H. W., Costa, J. H. C., von Keyserlingk, M. A. G. and Weary, D. M. 2019. Automatic weaning based on individual solid feed intake: Effects on behavior and performance of dairy calves. *J. Dairy Sci.* 102:5475-5491.

Bernardi, F., Fregonesi, J. A., Winckler, C., Veira, D. M., von Keyserlingk, M. A. G. and Weary, D. M. 2009. The stall-design paradox: neck rails increase lameness but improve udder and stall hygiene. *J. Dairy Sci.* 92(7):3074-3080.

Bertelsen, M. and Jensen, M. B. 2019. Does dairy calves' motivation for social play behaviour build up over time? *Appl. Anim. Behav. Sci.* 214:18-24.

Bolinger, D. J., Albright, J. L., Morrow-Tesch, J., Kenyon, S. J. and Cunningham, M. D. 1997. The effects of restraint using self-locking stanchions on dairy cows in relation to behavior, feed intake, physiological parameters, health, and milk yield. *J. Dairy Sci.* 80(10):2411-2417.

Bonk, S., Burfeind, O., Suthar, V. S. and Heuwieser, W. 2013. Technical note: evaluation of data loggers for measuring lying behavior in dairy calves. *J. Dairy Sci.* 96(5):3265-3271.

Borderas, T. F., De Passillé, A. M. and Rushen, J. 2008. Behavior of dairy calves after a low dose of bacterial endotoxin. *J. Anim. Sci.* 86(11):2920-2927.

Borderas, T. F., De Passillé, A. M. B. and Rushen, J. 2009. Temperature preferences and feed level of the newborn dairy calf. *Appl. Anim. Behav. Sci.* 120(1-2):56-61.

Brunsvold, R. E., Cramer, C.O. and Larsen, H. J. 1985. Behavior of dairy calves reared in hutches as affected by temperature. *Trans. ASAE* 28(4):1265-1268.

Buchli, C., Raselli, A., Bruckmaier, R. and Hillmann, E. 2017. Contact with cows during the young age increases social competence and lowers the cardiac stress reaction in dairy calves. *Appl. Anim. Behav. Sci.* 187:1-7.

Bučková, K., Špinka, M. and Hintze, S. 2019. Pair housing makes calves more optimistic. *Sci. Rep.* 9(1):20246.

Camiloti, T. V., Fregonesi, J. A., von Keyserlingk, M. A. G. and Weary, D. M. 2012. Short communication: effects of bedding quality on the lying behavior of dairy calves. *J. Dairy Sci.* 95(6):3380-3383.

Catanese, F., Obelar, M., Villalba, J. J. and Distel, R. A. 2013. The importance of diet choice on stress-related responses by lambs. *Appl. Anim. Behav. Sci.* 148(1-2):37-45.

Chen, J. M., Stull, C. L., Ledgerwood, D. N. and Tucker, C. B. 2017. Muddy conditions reduce hygiene and lying time in dairy cattle and increase time spent on concrete. *J. Dairy Sci.* 100(3):2090-2103.

Chua, B., Coenen, E., van, D. J. and Weary, D. M. 2002. Effects of pair versus individual housing on the behavior and performance of dairy calves. *J. Dairy Sci.* 85(2):360-364.

Coleman, D. A., Moss, B. R. and McCaskey, T. A. 1996. Supplemental shade for dairy calves reared in commercial calf hutches in a Southern climate. *J. Dairy Sci.* 79(11):2038-2043.

Costa, J. H. C., Daros, R. R., von Keyserlingk, M. A. G. and Weary, D. M. 2014. Complex social housing reduces food neophobia in dairy calves. *J. Dairy Sci.* 97(12):7804-7810.

Costa, J. H. C., von Keyserlingk, M. A. G. and Weary, D. M. 2016a. Invited review: effects of group housing of dairy calves on behavior, cognition, performance, and health. *J. Dairy Sci.* 99(4):2453-2467.

Costa, J. H. C., Adderley, N. A., Weary, D. M. and von Keyserlingk, M. A. G. 2016b. Short communication: effect of diet changes on sorting behavior of weaned dairy calves. *J. Dairy Sci.* 99(7):5635-5639.

Dado, R. G. and Allen, M. S. 1994. Variation in and relationships among feeding, chewing, and drinking variables for lactating dairy cows. *J. Dairy Sci.* 77(1):132-144.

Dahl, G. E., Skibiel, A. L. and Laporta, J. 2019. In utero heat stress programs reduced performance and health in calves. *Vet. Clin. North Am. Food Anim. Pract.* 35(2):343-353.

de Passillé, A. M. 2001. Sucking motivation and related problems in calves. *Appl. Anim. Behav. Sci.* 72(3):175-187.

de Passillé, A. M., Christopherson, R. and Rushen, J. 1993. Non-nutritive sucking by the calf and postprandial secretion of insulin, CCK and gastrin. *Physiol. Behav.* 54(6):1069-1073.

de Passillé, A. M., Sweeney, B. and Rushen, J. 2010. Cross-sucking and gradual weaning of dairy calves. *Appl. Anim. Behav. Sci.* 124(1-2):11-15.

de Passillé, A. M. and Rushen, J. 2016. Using automated feeders to wean calves fed large amounts of milk according to their ability to eat solid feed. *J. Dairy Sci.* 99(5):3578-3583.

De Paula Vieira, A., Guesdon, V., de Passille, A. M., von Keyserlingk, M. A. G. and Weary, D. M. 2008. Behavioral indicators of hunger in dairy calves. *Appl. Anim. Behav. Sci.* 109(2-4):180-189.

De Paula Vieira, A., von Keyserlingk, M. A. G. and Weary, D. M. 2010. Effects of pair versus single housing on performance and behavior of dairy calves before and after weaning from milk. *J. Dairy Sci.* 93(7):3079-3085.

De Paula Vieira, A., von Keyserlingk, M. A. G. and Weary, D. M. 2012. Presence of an older weaned companion influences feeding behavior and improves performance of dairy calves before and after weaning from milk. *J. Dairy Sci.* 95(6):3218-3224.

DeVries, T. J., Dohme, F. and Beauchemin, K. A. 2008. Repeated ruminal acidosis challenges in lactating dairy cows at high and low risk for developing acidosis: feed sorting. *J. Dairy Sci.* 91(10):3958-3967.

DeVries, T. J., Schwaiger, T., Beauchemin, K. A. and Penner, G. B. 2014. Impact of severity of ruminal acidosis on feed-sorting behaviour of beef cattle. *Anim. Prod. Sci.* 54(9):1238-1242.

Duve, L. R., Weary, D. M., Halekoh, U. and Jensen, M. B. 2012. The effects of social contact and milk allowance on responses to handling, play, and social behavior in young dairy calves. *J. Dairy Sci.* 95(11):6571-6581.

Eckert, E., Brown, H. E., Leslie, K. E., DeVries, T. J. and Steele, M. A. 2015. Weaning age affects growth, feed intake, gastrointestinal development, and behavior in Holstein calves fed an elevated plane of nutrition during the preweaning stage. *J. Dairy Sci.* 98(9):6315-6326.

Ede, T., von Keyserlingk, M. A. G. and Weary, D. M. 2018. Approach-aversion in calves following injections. *Sci. Rep.* 8(1):9443.

Fabris, T. F., Laporta, J., Skibiel, A. L., Corra, F. N., Sen, B. D., Wohlgemuth, S. E. and Dahl, G. E. 2019. Effect of heat stress during early, late, and entire dry period on dairy cattle. *J. Dairy Sci.* 102(6):5647-5656.

Færevik, G., Jensen, M. B. and Bøe, K. E. 2006. Dairy calves social preference and the significance of a companion animal during separation from the group. *Appl. Anim. Behav. Sci.* 99(3-4):205--221.

Færevik, G., Andersen, I. L., Jensen, M. B. and Bøe, K. E. 2007. Increased group size reduces conflicts and strengthens the preference for familiar group mates after regrouping of weaned dairy calves (Bos taurus). *Appl. Anim. Behav. Sci.* 108(3-4):215-228.

Favreau-Peigné, A., Baumont, R. and Ginane, C. 2013. Food sensory characteristics: their unconsidered roles in the feeding behavior of domestic ruminants. *Animal* 7(5):806-813.

Ferreira, F. C., Gennari, R. S., Dahl, G. E. and De Vries, A. 2016. Economic feasibility of cooling dry cows across the United States. *J. Dairy Sci.* 99(12):9931-9941.

Fisher, A. D., Stewart, M., Verkerk, G. A., Morrow, C. J. and Matthews, L. R. 2003. The effects of surface type on lying behaviour and stress responses of dairy cows during periodic weather-induced removal from pasture. *Appl. Anim. Behav. Sci.* 81(1):1-11.

Flower, F. C. and Weary, D. M. 2001. Effects of early separation on the dairy cow and calf: 2. Separation at 1 day and 2 weeks after birth. *Appl. Anim. Behav. Sci.* 70(4):275-284.

Fröberg, S., Aspegren-Guldorff, A., Olsson, I., Marin, B., Berg, C., Hernandez, C., Galina, C. S., Lidfors, L. and Svennersten-Sjaunja, K. 2007. Effect of restricted suckling on milk yield, milk composition and udder health in cows and behaviour and weight gain in calves, in dual-purpose cattle in the tropics. *Trop. Anim. Health Prod.* 39(1):71-81.

Fureix, C., Walker, M., Harper, L., Reynolds, K., Saldivia-Woo, A. and Mason, G. 2016. Stereotypic behaviour in standard non-enriched cages is alternative to depression-like responses in C57BL/6 mice. *Behav. Brain Res.* 305:186-190.

Gaillard, C., Meagher, R. K., von Keyserlingk, M. A. G. and Weary, D. M. 2014. Social housing improves dairy calves' performance in two cognitive tests. *PLoS ONE* 9(2):e90205.

Galindo, F. and Broom, D. M. 2000. The relationships between social behaviour of dairy cows and the occurrence of lameness in three herds. *Res. Vet. Sci.* 69(1):75-79.

Gingerich, K. N., V. Choulet and Miller-Cushon, E. K. 2020. Disbudding affects use of a shelter provided to group-housed dairy calves. *J. Dairy Sci.* https://doi.org/10.3168/jds.2020-18267.

Gonzalez-Jimenez, E. and Baxter, K. L. 1962. The metabolism and thermal regulation of calves in the first month of life. *Br. J. Nutr.* 16:199-212.

Greter, A. M., Miller-Cushon, E. K., McBride, B. W., Widowski, T. M., Duffield, T. F. and DeVries, T. J. 2015. Short communication: limit feeding affects behavior patterns and feed motivation of dairy heifers. *J. Dairy Sci.* 98(2):1248-1254.

Hagen, K. and Broom, D. M. 2004. Emotional reactions to learning in cattle. *Appl. Anim. Behav. Sci.* 85(3-4):203-213.

Haley, D. B., Rushen, J., Duncan, I. J., Widowski, T. M. and De Passillé, A. M. 1998. Effects of resistance to milk flow and the provision of hay on nonnutritive sucking by dairy calves. *J. Dairy Sci.* 81(8):2165-2172.

Hänninen, L., de Passillé, A. M. and Rushen, J. 2005. The effect of flooring type and social grouping on the rest and growth of dairy calves. *Appl. Anim. Behav. Sci.* 91(3-4):193-204.

Hill, T. M., Bateman, H. G., Aldrich, J. M. and Schlotterbeck, R. L. 2011. Comparisons of housing, bedding, and cooling options for dairy calves. *J. Dairy Sci.* 94(4):2138-2146.

Hixson, C. L., Krawczel, P. D., Caldwell, J. M. and Miller-Cushon, E. K. 2018. Behavioral changes in group-housed dairy calves infected with Mannheimia haemolytica. *J. Dairy Sci.* 101(11):10351-10360.

Holm, L., Jensen, M. B. and Jeppesen, L. L. 2002. Calves' motivation for access to two different types of social contact measured by operant conditioning. *Appl. Anim. Behav. Sci.* 79(3):175–194.
Horvath, K. C., Fernandez, M. and Miller-Cushon, E. K. 2017. The effect of feeding enrichment in the milk-feeding stage on the cognition of dairy calves in a T-maze. *Appl. Anim. Behav. Sci.* 187:8–14.
Horvath, K. C. and Miller-Cushon, E. K. 2017. The effect of milk-feeding method and hay provision on the development of feeding behavior and non-nutritive oral behavior of dairy calves. *J. Dairy Sci.* 100(5):3949–3957.
Horvath, K. C. and Miller-Cushon, E. K. 2018. Characterizing social behavior, activity, and associations between cognition and behavior upon social grouping of weaned dairy calves. *J. Dairy Sci.* 101(8):7287–7296.
Horvath, K. C. and Miller-Cushon, E. K. 2019. Characterizing grooming behavior patterns and the influence of brush access on the behavior of group-housed dairy calves. *J. Dairy Sci.* 102(4):3421–3430.
Horvath, K. C. and Miller-Cushon, E. K. 2020. Effects of hay provision and presentation on cognitive development in dairy calves. *PLoS ONE* 15(9): e0238038. https://doi.org/10.1371/journal.pone.0238038.
Horvath, K. C., Allen, A. N. and Miller-Cushon, E. K. 2020. Effects of access to stationary brushes and chopped hay on behavior and performance of individually housed dairy calves. *J. Dairy Sci.* 103:8421–8432.
Hosseinkhani, A., DeVries, T. J., Proudfoot, K. L., Valizadeh, R., Veira, D. M. and von Keyserlingk, M. A. G. 2008. The effects of feed bunk competition on the feed sorting behavior of close-up dry cows. *J. Dairy Sci.* 91(3): 1115–1121.
Huber, R., Baumung, R., Wurzinger, M., Semambo, D., Mwai, O. and Winckler, C. 2008. Grazing, social and comfort behaviour of Ankole and crossbred (Ankole x Holstein) heifers on pasture in south western Uganda. *Appl. Anim. Behav. Sci.* 112(3–4): 223–234.
Jasper, J. and Weary, D. M. 2002. Effects of ad libitum milk intake on dairy calves. *J. Dairy Sci.* 85(11):3054–3058.
Jensen, M. B., Vestergaard, K. S., Krohn, C. C. and Munksgaard, L. 1997. Effect of single versus group housing and space allowance on responses of calves during open-field tests. *Appl. Anim. Behav. Sci.* 54(2–3):109–121.
Jensen, M. B. and Kyhn, R. 2000. Play behaviour in group-housed dairy calves, the effect of space allowance. *Appl. Anim. Behav. Sci.* 67(1–2):35–46.
Jensen, M. B., Pedersen, L. J. and Munksgaard, L. 2005. The effect of reward duration on demand functions for rest in dairy heifers and lying requirements as measured by demand functions. *Appl. Anim. Behav. Sci.* 90(3–4):207–217.
Jensen, M. B. and Budde, M. 2006. The effects of milk feeding method and group size on feeding behavior and cross-sucking in group-housed dairy calves. *J. Dairy Sci.* 89(12):4778–4783.
Jensen, M. B., de Passillé, A. M., von Keyserlingk, M. A. G. and Rushen, J. 2008. A barrier can reduce competition over teats in pair-housed milk-fed calves. *J. Dairy Sci.* 91(4):1607–1613.
Jensen, M. B. 2011. The early behaviour of cow and calf in an individual calving pen. *Appl. Anim. Behav. Sci.* 134(3–4):92–99.

Jensen, M. B., Duve, L. R. and Weary, D. M. 2015. Pair housing and enhanced milk allowance increase play behavior and improve performance in dairy calves. *J. Dairy Sci.* 98(4):2568-2575.

Johnsen, J. F., de Passille, A. M., Mejdell, C. M., Bøe, K. E., Grøndahl, A. M., Beaver, A., Rushen, J. and Weary, D. M. 2015. The effect of nursing on the cow-calf bond. *Appl. Anim. Behav. Sci.* 163:50-57.

Johnsen, J. F., Zipp, K. A., Kälber, T., de Passillé, A. Md, Knierim, U., Barth, K. and Mejdell, C. M. 2016. Is rearing calves with the dam a feasible option for dairy farms?–current and future research. *Appl. Anim. Behav. Sci.* 181:1-11.

Jorgensen, M. W., Adams-Progar, A., de Passillé, A. M., Rushen, J., Godden, S. M., Chester-Jones, H. and Endres, M. I. 2017. Factors associated with dairy calf health in automated feeding systems in the Upper Midwest United States. *J. Dairy Sci.* 100(7):5675-5686.

Jung, J. and Lidfors, L. 2001. Effects of amount of milk, milk flow and access to a rubber teat on cross-sucking and non-nutritive sucking in dairy calves. *Appl. Anim. Behav. Sci.* 72(3):201-213.

Keil, N. M. and Langhans, W. 2001. The development of intersucking in dairy calves around weaning. *Appl. Anim. Behav. Sci.* 72(4):295-308.

Keunen, J. E., Plaizier, J. C., Kyriazakis, L., Duffield, T. F., Widowski, T. M., Lindinger, M. I. and McBride, B. W. 2002. Effects of a subacute ruminal acidosis model on the diet selection of dairy cows. *J. Dairy Sci.* 85(12):3304-3313.

Khan, M. A., Lee, H. J., Lee, W. S., Kim, H. S., Ki, K. S., Hur, T. Y., Suh, G. H., Kang, S. J. and Choi, Y. J. 2007. Structural growth, rumen development, and metabolic and immune responses of Holstein male calves fed milk through step-down and conventional methods. *J. Dairy Sci.* 90(7):3376-3387.

Khan, M. A., Weary, D. M. and von Keyserlingk, M. A. G. 2011a. Invited review: effects of milk ration on solid feed intake, weaning, and performance in dairy heifers. *J. Dairy Sci.* 94(3):1071-1081.

Khan, M. A., Weary, D. M. and von Keyserlingk, M. A. G. 2011b. Hay intake improves performance and rumen development of calves fed higher quantities of milk. *J. Dairy Sci.* 94(7):3547-3553.

Kilgour, R. J. 2012. In pursuit of "normal": a review of the behaviour of cattle at pasture. *Appl. Anim. Behav. Sci.* 138(1-2):1-11.

Kitts, B. L., Duncan, I. J. H., McBride, B. W. and DeVries, T. J. 2011. Effect of the provision of a low-nutritive feedstuff on the behavior of dairy heifers fed a high-concentrate ration in a limited amount. *J. Dairy Sci.* 94(2):940-950.

Kjæstad, H. P. and Myren, H. J. 2001a. Failure to use cubicles and concentrate dispenser by heifers after transfer from rearing accommodation to milking herd. *Acta Vet. Scand.* 42(1):171-180.

Kjæstad, H. P. and Myren, H. J. 2001b. Cubicle refusal in Norwegian dairy herds. *Acta Vet. Scand.* 42(1):181-187.

Kjæstad, H. P. and Simensen, E. 2001. Cubicle refusal and rearing accommodation as possible mastitis risk factors in cubicle-housed dairy heifers. *Acta Vet. Scand.* 42(1):123-130.

Kovács, L., Kézér, F. L., Póti, P., Boros, N. and Nagy, K. 2020. Short communication: upper critical temperature-humidity index for dairy calves based on physiological stress variables. *J. Dairy Sci.* 103(3):2707-2710.

Krohn, C. C. and Munksgaard, L. 1993. Behaviour of dairy cows kept in extensive (loose housing/pasture) or intensive (tie stall) environments II. Lying and lying-down behavior. *Appl. Anim. Behav. Sci.* 37(1):1-16.

Kyriazakis, I. and Oldham, J. D. 1993. Diet selection in sheep: the ability of growing lambs to select a diet that meets their crude protein (nitrogen x 6.25) requirements. *Br. J. Nutr.* 69(3):617-629.

Lago, A., McGuirk, S. M., Bennet, T. B., Cook, N. B. and Nordlund, K. V. 2006. Calf respiratory disease and pen microenvironments in naturally ventilated calf barns in winter. *J. Dairy Sci.* 89:4014-4025.

Laister, S., Stockinger, B., Regner, A. M., Zenger, K., Knierim, U. and Winckler, C. 2011. Social licking in dairy cattle–effects on heart rate in performers and receivers. *Appl. Anim. Behav. Sci.* 130(3-4):81-90.

Lecorps, B., Kappel, S., Weary, D. M. and von Keyserlingk, M. A. G. 2018. Dairy calves' personality traits predict social proximity and response to an emotional challenge. *Sci. Rep.* 8(1):16350.

Legrand, A., Schütz, K. E. and Tucker, C. B. 2011. Using water to cool cattle: behavioral and physiological changes associated with voluntary use of cow showers. *J. Dairy Sci.* 94(7):3376-3386.

Lewis, M. H., Tanimura, Y., Lee, L. W. and Bodfish, J. W. 2006. Animal models of restricted repetitive behavior in autism. *Behav. Brain Res.* 10:66-74.

Lidfors, L. M. and Isberg, L. 2003. Intersucking in dairy cattle - review and questionnaire. *Appl. Anim. Behav. Sci.* 80(3):207-231.

Lidfors, L. M., Moran, D., Jung, J., Jensen, P. and Castren, H. 1994. Behaviour at calving and choice of calving place in cattle kept in different environments. *Appl. Anim. Behav. Sci.* 42(1):11-28.

Lindstrom, T. and Redbo, I. 2000. Effect of feeding duration and rumen fill on behaviour in dairy cows. *Appl. Anim. Behav. Sci.* 70(2):83-97.

Leruste, H., Brscic, M., Cozzi, G., Kemp, B., Wolthuis-Fillerup, M., Lensink, B. J., Bokkers, E. A. M. and van Reenen, C. G. 2014. Prevalence and potential influencing factors of non-nutritive oral behaviors of veal calves on commercial farms. *J. Dairy Sci.* 97(11):7021-7030.

Loberg, J. and Lidfors, L. 2001. Effect of milkflow rate and presence of a floating nipple on abnormal sucking between dairy calves. *Appl. Anim. Behav. Sci.* 72(3): 189-199.

Marcé, C., Guatteo, R., Bareille, N. and Fourichon, C. 2010. Dairy calf housing systems across Europe and risk for calf infectious diseases. *Animal* 4(9):1588-1596.

Mason, G. J. 1991. Stereotypies: a critical review. *Anim. Behav.* 41(6):1015-1037.

McConnachie, E., Smid, A. M. C., Thompson, A. J., Weary, D. M., Gaworski, M. A. and von Keyserlingk, M. A. G. 2018. Cows are highly motivated to access a grooming substrate. *Biol. Lett.* 14(8):pii:20180303.

Meagher, R. K., Daros, R. R., Costa, J. H. C., von Keyserlingk, M. A. G., Hötzel, M. J. and Weary, D. M. 2015. Effects of degree and timing of social housing on reversal learning and response to novel objects in dairy calves. *PLoS ONE* 10(8):e0132828.

Meagher, R. K., Weary, D. M. and von Keyserlingk, M. A. G. 2017. Some like it varied: individual differences in preference for feed variety in dairy heifers. *Appl. Anim. Behav. Sci.* 195:8-14.

Meagher, R. K. 2019. Is boredom an animal welfare concern? *Animal Welfare* 28:21-32.

Meagher, R. K., Beaver, A., Weary, D. M. and von Keyserlingk, M. A. G. 2019. Invited review: a systematic review of the effects of prolonged cow-calf contact on behavior, welfare, and productivity. J. Dairy Sci. 102(7):5765-5783.

Metz, J. H. M. 1985. The reaction of cows to a short-term deprivation of lying. *Appl. Anim. Behav. Sci.* 13(4):301-307.

Miller-Cushon, E. K., Bergeron, R., Leslie, K. E. and DeVries, T. J. 2013a. Effect of milk feeding level on development of feeding behavior in dairy calves. *J. Dairy Sci.* 96(1):551-564.

Miller-Cushon, E. K., Bergeron, R., Leslie, K. E., Mason, G. J. and DeVries, T. J. 2013b. Effect of early exposure to different feed presentations on feed sorting of dairy calves. *J. Dairy Sci.* 96(7):4624-4633.

Miller-Cushon, E. K., Bergeron, R., Leslie, K. E., Mason, G. J. and DeVries, T. J. 2014a. Competition during the milk-feeding stage influences the development of feeding behavior of pair-housed dairy calves. *J. Dairy Sci.* 97(10):6450-6462.

Miller-Cushon, E. K., Montoro, C., Ipharraguerre, I. R. and Bach, A. 2014b. Dietary preference in dairy calves for feed ingredients high in energy and protein. *J. Dairy Sci.* 97(3):1634-1644.

Miller-Cushon, E. K., Terré, M., DeVries, T. J. and Bach, A. 2014c. The effect of palatability of protein source on dietary selection in dairy calves. *J. Dairy Sci.* 97(7):4444-4454.

Miller-Cushon, E. K. and DeVries, T. J. 2015. Invited review: development and expression of dairy calf feeding behaviour. *Can. J. Anim. Sci.* 95(3):341-350.

Miller-Cushon, E. K. and DeVries, T. J. 2016. Effect of social housing on the development of feeding behavior and social feeding preferences of dairy calves. *J. Dairy Sci.* 99(2):1406-1417.

Mintline, E. M., Stewart, M., Rogers, A. R., Cox, N. R., Verkerk, G. A., Stookey, J. M., Webster, J. R. and Tucker, C. B. 2013. Play behavior as an indicator of animal welfare: disbudding in dairy calves. *Appl. Anim. Behav. Sci.* 144(1-2):22-30.

Mirzaei, M., Dadkhah, N., Baghbanzadeh-Nobari, B., Agha-Tehrani, A., Eshraghi, M., Imani, M., Shiasi-Sardoabi, R. and Ghaffari, M. H. 2018. Effects of preweaning total plane of milk intake and weaning age on intake, growth performance, and blood metabolites of dairy calves. *J. Dairy Sci.* 101(5):4212-4220.

Montoro, C., Ipharraguerre, I. and Bach, A. 2011. Effect of flavoring a starter in a same manner as a milk replacer on intake and performance of calves. *Anim. Feed Sci. Tech.* 164(1-2):130-134.

Montoro, C. and Bach, A. 2012. Voluntary selection of starter feed ingredients offered separately to nursing calves. *Livest. Sci.* 149(1-2):62-69.

Moore, D. A., Duprau, J. L. and Wenz, J. R. 2012. Short communication: effects of dairy calf hutch elevation on heat reduction, carbon dioxide concentration, air circulation, and respiratory rates. *J. Dairy Sci.* 95(7):4050-4054.

Muller, C. J. C., Botha, J. A. and Smith, W. A. 1996. Effect of confinement area on production, physiological parameters and behaviour of Friesian cows during winter in a temperate climate. *S. Afr. J. Anim. Sci.* 26:1-5.

Munksgaard, L., Jensen, M. B., Pedersen, L. J., Hansen, S. W. and Matthews, L. 2005. Quantifying behavioral priorities: effects of time constraints on the behavior of dairy cows, Bos taurus. *Appl. Anim. Behav. Sci.* 92(1-2):3-14.

Neave, H. W., Costa, J. H. C., Weary, D. M. and von Keyserlingk, M. A. G. 2018. Personality is associated with feeding behavior and performance in dairy calves. *J. Dairy Sci.* 101(8):7437-7449.

Nicol, A. M. and Sharafeldin, M. A. 1975. Observations on the behaviour of single-suckled calves from birth to 120 days. *Proc. N.Z. Soc. Anim. Prod.* 35:221–230.

Nordlund, K. V. and Halbach, C. E. 2019. Calf barn design to optimize health and ease of management. *Vet. Clin. North Am. Food Anim. Pract.* 35(1):29–45.

O'Connell, J. M., Giller, P. S. and Meaney, W. J. 1993. Weanling training and cubicle usage as heifers. *Anim. Beh.* 37(3):185–195.

Owens, F. N., Secrist, D. S., Hill, W. J. and Gill, D. R. 1998. Acidosis in cattle: a review. *J. Anim. Sci.* 76(1):275–286.

Panivivat, R., Kegley, E. B., Pennington, J. A., Kellogg, D. W. and Krumpelma, S. L. 2004. Growth performance and health of dairy calves bedded with different types of materials. *J. Dairy Sci.* 87(11):3736–3745.

Parola, F., Hillmann, E., Schütz, K. E. and Tucker, C. B. 2012. Preferences for overhead sprinklers by naïve beef steers: test of two nozzle types. *Appl. Anim. Behav. Sci.* 137(1–2):13–22.

Pempek, J. A., Eastridge, M. L. and Proudfoot, K. L. 2017. The effect of a furnished individual hutch pre-weaning on calf behavior, response to novelty, and growth. *J. Dairy Sci.* 100(6):4807–4817.

Proudfoot, K. L., Jensen, M. B., Weary, D. M. and von Keyserlingk, M. A. G. 2014. Dairy cows seek isolation at calving and when ill. *J. Dairy Sci.* 97(5):2731–2739.

Provenza, F. D. and Balph, D. F. 1987. Diet learning by domestic ruminants: theory, evidence and practical implications. *Appl. Anim. Behav. Sci.* 18(3–4):211–232.

Redbo, I. 1990. Changes in duration and frequency of stereotypies and their adjoining behaviours in heifers, before, during and after the grazing period. *Appl. Anim. Behav. Sci.* 26(1–2):57–67.

Redbo, I. and Nordblad, A. 1997. Stereotypies in heifers are affected by feeding regime. *Appl. Anim. Behav. Sci.* 53(3):193–202.

Reinhardt, V. and Reinhardt, A. 1981. Cohesive relationships in a cattle herd (Bos indicus). *Behaviour* 77(3):121–150.

Richter, S. H. and Hintze, S. 2019. From the individual to the population – and back again? Emphasising the role of the individual in animal welfare science. *Appl. Anim. Behav. Sci.* 212:1–8.

Roth, B. A., Hillman, E., Stauffacher, M. and Keil, N. M. 2008. Improved weaning reduces cross-sucking and may improve weight gain in dairy calves. *Appl. Anim. Behav. Sci.* 111(3–4):251–261.

Roth, B. A., Barth, K., Gygax, L. and Hillmann, E. 2009. Influence of artificial vs. mother-bonded rearing on sucking behaviour, health and weight gain in calves. *Appl. Anim. Behav. Sci.* 119(3–4):143–150.

Roura, E., Humphrey, B., Tedo, G. and Ipharraguerre, I. 2008. Unfolding the codes of short-term feed appetence in farm and companion animals. A comparative oronasal nutrient sensing biology review. *Can. J. Anim. Sci.* 88(4):535–558.

Sato, S., Wood-Gush, D. G. M. and G, W. 1987. Observations on creche behaviour in suckler calves. *Behav. Processes* 15(2–3):333–343.

Scott, L. L. and Provenza, F. D. 1998. Variety of foods and flavors affects selection of foraging location by sheep. *Appl. Anim. Behav. Sci.* 61(2): 113–122.

Spain, J. N. and Spiers, D. E. 1996. Effects of supplemental shade on thermoregulatory response of calves to heat challenge in a hutch environment. *J. Dairy Sci.* 79(4):639–646.

Špinka, M. 2019. Animal agency, animal awareness and animal welfare. *Anim. Welf.* 28(1):11–20.

Sutherland, M. A., Stewart, M. and Schütz, K. E. 2013. Effects of two substrate types on the behaviour, cleanliness and thermoregulation of dairy calves. *Appl. Anim. Behav. Sci.* 147(1-2):19–27.

Sutherland, M. A., Worth, G. M., Schütz, K. E. and Stewart, M. 2014. Rearing substrate and space allowance influences locomotor play behaviour of dairy calves in an arena test. *Appl. Anim. Behav. Sci.* 154:8–14.

Sutherland, M. A., Worth, G. M., Cameron, C., Ross, C. M. and Rapp, D. 2017. Health, physiology, and behavior of dairy calves reared on 4 different substrates. *J. Dairy Sci.* 100(3):2148–2156.

Sweeney, B. C., Rushen, J. P., Weary, D. M. and de Passillé, A. M. 2010. Duration of weaning, starter intake, and weight gain of dairy calves fed large amounts of milk. *J. Dairy Sci.* 93(1):148–152.

Thomas, T. J., Weary, D. M. and Appleby, M. C. 2001. Newborn and 5-week-old calves vocalize in response to milk deprivation. *Appl. Anim. Behav. Sci.* 74(3):165–173.

USDA 2010. *Facility Characteristics and Cow Comfort on U.S. Dairy Operations, 2007*. UDSA-Animal and Plant Health Inspection Service-Veterinary Services, Centers for Epidemiology and Animal Health, Fort Collins, CO.

USDA 2016. *Dairy Cattle Management Practices in the United States, 2014*. UDSA-Animal and Plant Health Inspection Service-Veterinary Services, Centers for Epidemiology and Animal Health, Fort Collins, CO.

Valníčková, B., Stěhulová, I., Šárová, R. and Špinka, M. 2015. The effect of age at separation from the dam and presence of social companions on play behavior and weight gain in dairy calves. *J. Dairy Sci.* 98(8):5545–5556.

Van Os, J. M. C., Mintline, E. M., DeVries, T. J. and Tucker, C. B. 2018. Domestic cattle (Bos taurus taurus) are motivated to obtain forage and demonstrate contrafreeloading. *PLoS ONE* 13(3):e0193109.

Van Os, J. M. C., Goldstein, S. A., Weary, D. M. and von Keyserlingk, M. A. G. 2019. Brush use by dairy heifers. Proceedings of the 53rd Congress of the International Society for Applied Ethology, Bergen, Norway.

Vasseur, E., Borderas, F., Cue, R. I., Lefebvre, D., Pellerin, D., Rushen, J., Wade, K. M. and de Passillé, A. M. 2010. A survey of dairy calf management practices in Canada that affect animal welfare. *J. Dairy Sci.* 93(3):1307–1315.

Vaughan, A., Miguel-Pacheco, G. G., de Passillé, A. M. and Rushen, J. 2016. Reciprocated cross sucking between dairy calves after weaning off milk does not appear to negatively affect udder health or production. *J. Dairy Sci.* 99(7):5596–5603.

Ventura, B. A., von Keyserlingk, M. A., Schuppli, C. A. and Weary, D. M. 2013. Views on contentious practices in dairy farming: the case of early cow-calf separation. *J. Dairy Sci.* 96(9):6105–6116.

Veissier, I., de Passillé, A. M., Després, G., Rushen, J., Charpentier, I., Ramirez de la Fe, A. R. and Pradel, P. 2002. Does nutritive and non-nutritive sucking reduce other oral behaviors and stimulate rest in calves? *J. Anim. Sci.* 80(10):2574–2587.

von Keyserlingk, M. A. G., Cunha, G. E., Fregonesi, J. A. and Weary, D. M. 2011. Introducing heifers to freestall housing. *J. Dairy Sci.* 94(4):1900–1907.

Weary, D. M., Jasper, J. and Hötzel, M. J. 2008. Understanding weaning distress. *Appl. Anim. Behav. Sci.* 110(1-2):24–41.

Webb, L. E., Bokkers, E. A. M., Engel, B., Gerrits, W. J. J., Berends, H. and van Reenen, C. G. 2012. Behaviour and welfare of veal calves fed different amounts of solid feed supplemented to a milk replacer ration adjusted for similar growth. *Appl. Anim. Behav. Sci.* 136(2–4):108–116.

Webb, L. E., Engel, B., Berends, H., van Reenen, C. G., Walter, W. J. J., de Boer, I. J. M. and Bokkers, E. A. M. 2014a. What do calves choose to eat and how do preferences affect behaviour? *Appl. Anim. Behav. Sci.* 161:7–19.

Webb, L. E., Jensen, M. B., Engel, B., van Reenen, C. G., Gerrits, W. J. J., de Boer, I. J. M. and Bokkers, E. A. M. 2014b. Chopped or long roughage: what do calves prefer? Using cross point analysis of double demand functions. *PLoS ONE* 9(2):e88778.

Webb, L. E., Engel, B., van Reenen, K. and Bokkers, E. A. M. 2017. Barren diets increase wakeful inactivity in calves. *Appl. Anim. Behav. Sci.* 197:9–14.

Whalin, L., Weary, D. M. and von Keyserlingk, M. A. G. 2018. Short communication: pair housing dairy calves in modified calf hutches. *J. Dairy Sci.* 101(6):5428–5433.

Winder, C. B., LeBlanc, S. J., Haley, D. B., Lissemore, K. D., Godkin, M. A. and Duffield, T. F. 2017. Clinical trial of local anesthetic protocols for acute pain associated with caustic paste disbudding in dairy calves. *J. Dairy Sci.* 100(8):6429–6441.

Wormsbecher, L., Bergeron, R., Haley, D., de Passillé, A. M., Rushen, J. and Vasseur, E. 2017. A method of outdoor housing dairy calves in pairs using individual calf hutches. *J. Dairy Sci.* 100(9):7493–7506.

Worth, G. M., Schütz, K. E., Stewart, M., Cave, V. M., Foster, M. and Sutherland, M. A. 2015. Dairy calves' preference for rearing substrate. *Appl. Anim. Behav. Sci.* 168:1–9.

Zobel, G., Neave, H. W., Henderson, H. V. and Webster, J. R. 2017. Calves use an automated brush and a hanging rope when pair-housed. *Animals* 7(11):84.

Chapter 9

Advances in understanding the needs and improving the welfare of transition dairy cows

Julia Lomb, University of British Columbia, Canada; and Kathryn L. Proudfoot, University of Prince Edward Island, Canada

1 Introduction

A dairy cow gives birth to a calf approximately once a year. Calving coincides with a period of tremendous physiological change as the cow transitions from a non-lactating, pregnant state to the lactating state. The time around parturition is often referred to as the 'transition' period, beginning three weeks before and ending three weeks after giving birth (Grummer, 1995; Drackley, 1999). The physical and physiological demands placed on the cow during transition are paired with changes in the cow's environment, such as separation from the calf, moving to a new social group and adapting to a new diet. For the cow giving birth for the first time, this also includes introduction to the milking parlor, as well as potentially comingling with older cows for the first time. Perhaps because of this multitude of challenges the transition period is a time when dairy cows are at increased risk of becoming ill (LeBlanc et al., 2006), also resulting in high rates of mortality and involuntary culling during the first 60 days of lactation (Dechow and Goodling, 2008). Not surprisingly, some have argued that transition cow disease is one of the greatest welfare challenges facing the dairy industry (von Keyserlingk et al., 2009).

http://dx.doi.org/10.19103/AS.2020.0084.11

Much of the research to date on transition dairy cows has focused on improvements in health. However, animal welfare encompasses more than an animal's physical health. This chapter will address the research to date measuring and assessing the welfare of transition cows using a framework described by Fraser et al. (1997), including animals' physical health ('biological functioning'), their emotional states ('affective state'), and their ability to live a reasonably natural life ('natural living').

2 Biological functioning

To date, research emphasis on transition cow welfare has focused mainly on biological functioning (see von Keyserlingk et al., 2009); this also corresponds well with the views of farmers who often cite health and production as the primary factors associated with care of dairy cattle (Te Velde et al., 2002; Lassen et al., 2006; Verbeke, 2009). Improving transition cow health may come about by prevention of disease and by early identification of ill cows allowing for timely intervention.

Cows are at high risk of both infectious and metabolic disease in the 3 weeks after calving; some argue that about 30–50% of all cows become ill during this time (LeBlanc, 2010). Among the most frequently reported metabolic diseases are subclinical and clinical hypocalcemia (Goff, 2008; Mulligan and Doherty, 2008; Reinhardt et al., 2011) and subclinical and clinical ketosis (Suthar et al., 2013; Tatone et al., 2017). Common infectious diseases observed in the weeks following calving include metritis (Giuliodori et al., 2013; Vergara et al., 2014; Pohl et al., 2016) and mastitis (De Vliegher et al., 2012). In addition to clinical disease, a high proportion of transition dairy cows show at least one day of fever during the first 10 days of lactation (Wenz et al., 2011: 14% in primiparous cows; 15% in multiparous cows; Suthar et al., 2012: up to 46.8% during high ambient temperatures), possibly in response to an undiagnosed infection or tissue damage.

Energy demands on the cow are high during the transition period, with the main sources for energy expense being for fetal growth during late gestation, followed by colostrogenesis and lactogenesis after parturition (Bell, 1995). Every mammal goes through a period of negative energy balance (NEB) after giving birth where metabolic demands associated with lactation exceed dietary energy intake. The challenge is when this period is extended (Grummer et al., 2004) and the mobilization of body fat, to support gluconeogenesis in the liver (Herdt, 2000), results in metabolic dysfunction. This metabolic challenge is especially problematic for high-yielding dairy cows (Ingvartsen et al., 2003; Oltenacu and Broom, 2010).

The high risk for these diseases is in part impacted by the high metabolic demand of lactation, as metabolic stress has been found to impair the functioning of other physiological processes, such as the immune (Zerbe et al., 2000; Hammon et al., 2006) and reproductive system (Chagas et al., 2007; Aleri et al., 2016). Consequently, incidences of metabolic and inflammatory disease have been reported higher in cows that were metabolically challenged, as evidenced by elevated metabolic blood parameters (Ruprechter et al., 2018) and by loss of body condition during the dry period (Chebel et al., 2018).

The prevention and early identification of diseases during the transition period is key to reducing this animal welfare concern. Preventing disease should include an approach that considers the animal's environment, their diet, their exposure to pathogens, as well as their ability to defend against the illness. Proper nutritional management plays a key role in disease prevention for transition cows. For example, a diet that does not provide a transition cow with sufficient energy increases the risk for a prolonged negative energy balance (Grummer, 2008) and metabolic disease (Overton and Waldron, 2004). Further, diet composition may provide the cow with sub-optimal levels of nutrients important for immune functioning (Sordillo, 2016). Lastly, energy-dense diets carry the risk for lowering ruminal pH and result in sub-acute ruminal acidosis (SARA; Humer et al., 2018).

Social stressors in the cow's environment should also be avoided where possible, as these can result in changes in behavior and immunity that may increase a cow's risk of illness. Research using humans and laboratory animals has shown linkages between social stress and disease risk (e.g. Glaser et al., 1992; Segerstrom and Miller, 2004; Hennessy et al., 2010), but to date few studies have addressed this issue in dairy cattle (reviewed by Proudfoot and Habing, 2015; Chebel et al., 2016). Like other species, common sources of social stress in transition cows likely include overcrowding (e.g. overstocking the lying or feeding areas), social instability (e.g. regrouping and moving animals between pens), and social status (e.g. the individual's social rank within the group). Throughout lactation, cows are typically grouped depending on their lactational stage and nutritional needs. In the peripartum period, pen moves are frequent and cows are forced to regularly comingle with unfamiliar conspecifics (see Cook and Nordlund, 2004). In a recent survey of dairy farmers in the United Kingdom, 84% reported that they housed dry cows in dynamic social groups with regular introductions of new cows to the group (Fujiwara et al., 2018). Moreover, overstocking is not uncommon in confinement dairy herds (Endres and Espejo, 2010; Sova et al., 2013), and 14% of UK farmers reported that not all pre-partum cows in their herd could access feed at the same time (Fujiwara et al., 2018).

A majority of the research on social stressors in transition cows has focused on the impact of overstocking on cow behavior. For example, transition cows at higher stocking density have more agonistic interactions at the feeding area (Proudfoot et al., 2009b; Lobeck-Luchterhand et al., 2015; Miltenburg et al., 2018), and have shorter lying times (Lobeck-Luchterhand et al., 2015; Miltenburg et al., 2018) compared with those at lower densities. Proudfoot et al. (2009b) also reported that cows that were displaced most often ate at a faster rate. Schirmann et al. (2011) found that cows regrouped into a new pen showed greater behavioral changes than cows that remained in the pen; cows that were introduced decreased feed intake and were replaced more often by other cows at the feeder. Some effects of regrouping may be different depending on the group size the moved cow is introduced to (Jensen and Proudfoot, 2017); cows that were moved to a group with fewer cows showed less agonistic interactions than cows introduced to a larger group, while cows of both groups increased feeding and lying behaviors over time. Jersey cows that were housed in either stable groups (control) or groups subjected to weekly regrouping also showed more displacements at the feed bunk in the regrouped pens (Lobeck-Luchterhand et al., 2014).

Some studies have attempted to directly link social stressors to physiological changes and disease risk, but results are inconsistent. For example, Huzzey et al. (2012b) found that overstocking cows at the feeding and lying areas during the dry period resulted in higher non-esterified fatty acids (NEFA) concentrations and a tendency for higher fecal cortisol concentrations compared to those that were not overstocked. Similarly, Proudfoot et al. (2018) created an unpredictable and overstocked feeding environment before calving, and found that cows had higher NEFA, tumor necrosis factor (TNF)-α and were at a higher risk of cytological endometritis compared to those in a more predictable social environment. In contrast, another study (Miltenburg et al., 2018) found that cows in understocked (80% stocking density) and overstocked (120% stocking density) groups throughout transition had only moderate differences in some metabolic and inflammatory markers. Moreover, there were no differences in metabolic or inflammatory markers or health outcomes between pre-partum Jersey cows housed at different stocking densities (80% vs. 100%; Silva et al., 2016) or in stable versus dynamic groups (Silva et al., 2013a,b).

Individual differences in disease risk are also important to consider. For example, there is some evidence that cows of low social rank may be more affected by a challenging environment. Lactating cows with a lower rank at the feed bunk appear to benefit most from less competitive environments at the feeder (DeVries et al., 2004; Huzzey et al., 2006). When assessed in the dry period, low-ranking cows appeared to have a higher stress response, measured as fecal cortisol, than higher-ranking cows (Huzzey et al., 2012a). In addition, primiparous but not multiparous cows housed in mixed-parity groups showed increased cortisol levels in conditions with limited access to the feed bunk (Huzzey et al., 2012b). First parity cows are likely to be of lower social rank when housed together with multiparous cows (González et al., 2003).

In addition to disease prevention, early detection of disease can aid in rapid treatment and reduced animal welfare challenges with sick animals. Traditionally, research and industry have sought to prevent disease by optimizing the physiological adaptation of the cow to transition, for example, by using nutritional strategies that improve metabolic health and immune function (e.g. Grummer, 2008; Zebeli et al., 2015), and by using biological markers to identify cows with and at risk for metabolic and inflammatory disease (e.g. Ospina et al., 2010; Shin et al., 2018). Despite these efforts, disease rates in transition cows remain relatively high. A newer approach to identify preventative management strategies and to detect ill cows early is the use of behavior and behavior change (Weary et al., 2009; Sepúlveda-Varas et al., 2013). This approach is becoming increasingly relevant, as technologies that allow for better monitoring of behavior and behavior change are becoming widely available on farms.

Early identification of disease can be improved using holistic approach that includes the animal's physical signs of illness as well as behavioral indicators. 'Sickness behaviors' are those behaviors that change when an animal becomes ill, and commonly include anorexia (a reduction in feed intake), adipsia (a reduction in water intake), depression and reduced activity, increased resting time, and avoidance of social interactions (Hart, 1988; Dantzer and Kelley, 2007). These behaviors, once thought of as an outcome of the individual's debilitation, are now considered a strategy that supports the mounting of a fever and overall immunological defense system of the organism (Hart, 1988). The magnitude to which sickness behaviors are expressed has been shown to vary with specific environmental and individual factors, for example social context (Hennessy et al., 2014) and distress (Avitsur et al., 2013).

While it is widely agreed upon that fever and sickness behaviors are beneficial to animals, the link between these behaviors and health is not clear. Fever has been shown to enhance functioning of the innate immune system rather than inhibiting pathogen growth directly (Blatteis, 2003). Sickness behaviors were first thought to primarily support the energetically costly fever response by conserving energy (i.e. by increasing resting time) and reduce support for pathogen proliferation (i.e. by inducing anorexia and subsequent lowering of blood iron concentrations; Hart, 1988). Others now argue that a primary function of sickness behaviors may be to limit pathogen distribution within a population (Shakhar and Shakhar, 2015).

Much of the research on sickness behaviors in cows has focused on feeding behavior. For example, early lactation cows infused with *E. coli* in one quarter decreased their feeding time for 24 h after infusion (Fogsgaard et al., 2012). In another study, cows which received intra-mammary *E. coli* LPS had decreased DMI (1 kg/d) on the day of infusion, but no overall decrease in daily rumination (Fitzpatrick et al., 2013). Cows with naturally occurring mastitis had lower feed intake, lower feeding rates, and also shorter lying times when compared with healthy cows even 10 days before mastitis was diagnosed and treated (Fogsgaard et al., 2015). Cows with naturally occurring metritis had lower feed intake compared to healthy cows in the first 2 (Schirmann et al., 2016) and 3 weeks postpartum (Huzzey et al., 2007) and spent less time at the feeder during the first 3 weeks after calving (Urton et al., 2005; Huzzey et al., 2007). It is noteworthy that some of these behavioral differences between healthy and metritic cows were already evident during the prepartum period (Huzzey et al., 2007), suggesting that some of these behaviors should be considered risk factors for illness.

Changes in lying time have also been measured as a result of sickness, but findings suggest the cow's response is dependent on the disease. Compared with control cows, primiparous animals with metritis had higher lying times in the days after diagnosis (Barragan et al., 2018). Similarly, primiparous cows diagnosed with a combination of multiple health disorders (i.e. metritis, retained placenta or mastitis) had longer lying times in the days after parturition, and tended to have longer lying bouts (Sepúlveda-Varas et al., 2014). However, cows with mastitis were consistently reported to increase standing times (e.g. Cyples et al., 2012; Fogsgaard et al., 2015). The increase in standing time for cows with mastitis is thought to be driven by pain, as cows may be more willing to stand than put pressure on their infected udder by lying down.

To date, changes in social behaviors as a sign of sickness have garnered less attention in dairy cows, in spite of being an integral part of the sickness response (Shakhar and Shakhar, 2015). One study found that cows with a fever and clinical sign of infectious disease in the few days after calving spent more time using a secluded area of pen compared to healthy cows (Proudfoot et al., 2014b). More research is encouraged to determine the usefulness of social behaviors as early indicators of illness.

Research focusing on the biological functioning of cows during transition has been diverse and combined knowledge of animal physiology, nutrition, social stressors, and behavior. This work has led to many advancements in our knowledge of transition cow diseases, as well as approaches for disease prevention and early disease detection. Key take homes from this research are that the animal's whole environment, including the physical, nutritional, and social environment, should be considered when developing disease prevention strategies. Moreover, early disease detection should incorporate behavior in addition to more traditional signs of illness.

3 Affective states

Parturition in any mammal is considered painful (Mainau and Manteca, 2011). Pain is defined by the International Association for the Study of Pain as an 'unpleasant sensory and emotional experience normally associated with tissue damage or described in terms of such damage.' In humans, labor pain is measured using subjective pain scales whereby a majority of women report moderate, intense, or very intense labor pain. In cows and other nonhuman animals, pain is often measured using behavioral indicators (Weary et al., 2006). Some general behaviors are associated with pain, including reluctance to eat, asocial behavior, avoidance behavior, increased attention to the painful area, and reductions in normal activity. Other behaviors are more specific to the source of the pain; for example, calves respond to painful disbudding by shaking their heads (Faulkner and Weary, 2000).

Transition dairy cows show behavioral changes in the days and hours before parturition that may indicate labor pain. Among the most frequently observed behaviors are an increase in the number of transitions between standing and lying before calving (Miedema et al., 2011a,b; Jensen, 2012) and shorter lying times on the days and hours before parturition (Huzzey et al., 2005; Jensen, 2012) combined with increased activity (i.e. average acceleration per second, measured by an electronic data logger attached to the leg; Jensen, 2012). Cows have also been reported to pay more attention to their abdomen as labor approaches (Jensen, 2012).

Cows that experience dystocia (i.e. parturition with difficulties requiring human assistance) are thought by veterinarians to experience high levels of pain (Remnant et al., 2017). It has been estimated that between 13.7% and 29% of multiparous cows and 22.6% and 51% of primiparous cows experience dystocia (Mee, 2008). Cows with dystocia are at higher risk of disease (Peeler et al., 1994) and have shown behavioral indicators of pain. For example, cows with dystocia have shown more marked changes in behavior before parturition, including higher number of transitions from standing to lying, as well as lower feed and water intake in the few days before calving (Proudfoot et al., 2009a). A recent study also reported that cows with longer calving times and more abdominal contractions, likely resulting in a higher degree of pain, preferred to calve in the space with the highest isolation level, while cows with eutocia did not show a clear preference (Rørvang et al., 2017).

Cows with disease after calving also show behavioral indicators of pain. For example, Stojkov et al. (2015) found that cows with metritis show more pronounced back arch and increased heart rate variability in response to palpation of the uterus compared to those without metritis, suggesting that they were feeling visceral pain. Cows with metritis also had elevated concentrations of Substance P (Barragan et al., 2018), a physiological marker associated with pain. Similarly, cows with mastitis have lower feed intake and spend more time standing in avoidance of pressure placed on the mammary gland (Fitzpatrick et al., 2013; Siivonen et al., 2011; Medrano-Galarza et al., 2012). Moreover, farmers perceive both mastitis and metritis to be painful for cows (Thomsen et al., 2012).

In general, research on affective states in transition dairy cows is lacking. Research to date has focused on determining behavioral indicators of pain in cows experiencing normal and assisted labor. In addition, some research has identified behavioral indicators of pain in sick cows after calving. More research focusing on other affective states (e.g. depression, boredom, frustration, pleasure, etc.) is encouraged during this sensitive time period around parturition.

4 Natural living

Understanding the 'natural' behavior of domesticated animals can be difficult, as they are usually housed in environments unlike the natural habitats of their ancestors. One approach to measuring the natural behavior of domesticated species, such as dairy cows, is to move them from their unnatural environments into more natural environments and record their behavioral response. Researchers have used this approach with transition dairy cows, with a specific focus on the few days just before and after giving birth. For example, Lidfors et al. (1994) kept dairy cows on 50 acres of variable terrain and measured the behavior of cows as calving approached as well as the characteristics of a calving site. On the day of calving, cows were found to be more than 15 m from their nearest neighbor compared to the 4 previous days, and calved in areas with high altitude, tree cover, and branches overhead, or a man-made shelter with water, minerals, and hay. These authors and others (e.g. Rørvang et al., 2018b) have speculated that this isolation-seeking behavior before parturition may be both an anti-predator strategy, and may help facilitate the bond between the dam and her calf without disturbance from other cows.

The desire to isolate at calving is likely impacted by a number of factors, including the animal's age, their previous experience, the weather, and their access to resources. For example, Edwards et al. (2020) provided pre-parturient Holstein primiparous and multiparous cows with access to a covered barn and a 2.1 ha area of pasture, including an 'open' area of pasture with no natural forage cover, and a smaller area of pasture with tall grass and tree cover ('natural forage area'). Researchers found that primiparous cows giving birth for the first time were more likely to use the area of natural forage on pasture to calve, whereas older cows were more likely to use the barn to calve. Moreover, animals were more likely to use the area of natural forage when the temperature and humidity were high (THI > 68 and ≤ 79).

Researchers have also found that indoor-housed cows will isolate when given the opportunity, but this behavior is impacted by their social environment and time of day. For example, Proudfoot et al. (2014a) found that cows kept in an individual maternity pen with access to a large (2.4 m x 6.1 m) plywood structure ('shelter') are more likely to calve in the structure compared to a more open, exposed area of the pen (61% vs 39%). Cows began to use the shelter more often in the 8 h before calving, potentially coinciding with the first stage of labor. Cows were most likely to use the shelter during the daytime; 81% of cows that calved during daylight hours used the shelter to give birth.

Providing cows with the ability to isolate may be more challenging when kept in a group pen at calving. For example, Jensen and Rørvang (2018) found that only 10% of cows calved in a man-made hiding space within a group pen when cows had to compete for access to the space (2 cows: 1 hide). However, using a similar hiding space, Rørvang et al. (2018a) found that approximately 50% of cows calved in the hide when there was no competition (1 cow: 1 hide). Thus, when providing access to a resource for cows to isolate during calving producers should ensure that there is no competition for the space.

Research on the natural behavior of transition cows has focused mainly on the period just before and after calving. More research is encouraged to determine the natural behavior of cows at other time periods to allow for the developments of more naturalistic environments for dairy cows throughout the transition period.

5 Case study: Maternity pen design and management

To improve animal welfare, it is important that all three elements addressed in this chapter (biological functioning, affective state and natural living) be addressed. A specific example of the link between these three animal welfare concepts is a study conducted by Proudfoot et al. (2013). The objectives of the study were to determine when cows should be moved from a group pre-calving pen into an individual maternity pen. A caretaker moved each cow at one of the 3 time points: before labor, during early labor (signs of Stage 1 labor), and during late labor (signs of late Stage 1/early Stage 2 labor). Researchers measured the impact of these treatments on lying time during the second stage of labor, labor length, and generalized inflammatory response after birth (haptoglobin). Cows that were moved during late labor were found to have lower lying time, longer labor, and higher inflammation after birth. These results indicate that human disturbance during a late stage of labor can impact all three elements of animal behavior: her natural behavior (lying time), her affective states (pain associated with prolonged labor), and her biological functioning (inflammation after calving).

Including all three concerns about animal welfare allowed researchers to cater their recommendations to help improve the overall welfare of transition cows. For example, authors recommended producers that currently move the cow into an individual pen before calving to move cows before they are too far along in labor (e.g. at early signs of labor when her desire to hide may begin).

Authors also recommend that producers try not to disturb cows as much as possible while they are undergoing labor.

In addition to these recommendations, other studies described throughout this chapter can provide insight into the proper design and management of maternity areas. For example, to help reduce disease after calving, maternity areas should be cleaned out regularly (Elbers et al., 1998). To improve comfort during painful birth, the maternity area should also be well-bedded with straw and a layer of sand for extra drainage and added comfort (Campler et al., 2014). Efforts should also be made to allow for the cow to isolate if she chooses to in either an individual or a group maternity area (e.g. Proudfoot et al., 2014a,b; Rørvang et al., 2017). Lastly, the use of cameras in maternity pens is highly recommended, as these can allow dairy producers to keep close watch in their parturient cows without needing to disturb the cow.

6 Conclusion and future trends

To summarize, the transition cow must undergo a number of stressors before and after giving birth. These stressors, including pain during labor and a high risk of disease after calving contribute to an overall increase in animal welfare challenges compared to other stages of lactation. A majority of the research to date on the transition dairy cow has focused on her biological functioning and health. More research is encouraged to capture outcomes related to affective states and natural living during this vulnerable time period.

Research is also encouraged to incorporate technology and precision dairy farming with the research on sickness behaviors discussed here. Some research has already identified the usefulness of activity monitors and rumination sensors to help detect illness in dairy cows after calving (reviewed by Eckelkamp, 2019). There remains opportunity to continue this line of research, with the addition of other behaviors (e.g. social behavior) using increasingly low-cost and available technologies. Data collected by automated milking systems may also provide a useful resource to identify ill cows (King et al., 2018).

A future trend may be the investigation and use of analgesics for cows during labor, especially those with prolonged or assisted births. Analgesics may also be useful alongside traditional antimicrobial therapies for diseases after calving (e.g. metritis and mastitis). Some research has assessed the impact of non-steroidal anti-inflammatory drugs (NSAID), alone or in combination with antimicrobials, on metritis, but results have varied (Amiridis et al., 2001; Drillich et al., 2007; Pohl et al., 2016). It should be noted that these studies focused only on health and reproductive outcomes, not on the impact on the cow's behavioral response. Thus, future work is encouraged to measure the impact of analgesics such as NSAIDs on pain-specific behaviors in transition dairy cows.

Another research trend is measuring the impact of separation from the calf on cow and calf behavior, health, and affect. Although still in its infancy research in this area is showing that both the cow and calf may benefit from prolonged contact after birth, with no increases in health outcomes (reviewed by Beaver et al., 2019 and Meagher et al., 2019). More research is encouraged to determine the possible benefits of some cow-calf contact on affective states experienced by both the dam and her calf.

Researchers are also encouraged to partner with agricultural engineers to design creative ways for cows to separate from penmates at calving. Pens that allow cows to easily self-segregate from a group if they choose would prevent any negative impacts of human disturbance during labor. The impacts of space allowance and regrouping in group maternity pens is also an area of research worth further investigation, as these may impact her ability to isolate and experience a normal birth.

7 Where to look for further information

More information can be found through the Animal Welfare Program at the University of British Columbia, as well as the Sir James Dunn Animal Welfare Centre at the University of Prince Edward Island.

8 References

Aleri, J. W., Hine, B. C., Pyman, M. F., Mansell, P. D., Wales, W. J., Mallard, B. and Fisher, A. D. (2016). Periparturient immunosuppression and strategies to improve dairy cow health during the periparturient period, *Research in Veterinary Science* 108, 8–17. doi: 10.1016/j.rvsc.2016.07.007.

Amiridis, G. S., Leontides, L., Tassos, E., Kostoulas, P. and Fthenakis, G. C. (2001). Flunixin meglumine accelerates uterine involution and shortens the calving-to-first-oestrus interval in cows with puerperal metritis, *Journal of Veterinary Pharmacology and Therapeutics* 24(5), 365–367. doi: 10.1046/j.1365-2885.2001.00358.x.

Avitsur, R., Maayan, R. and Weizman, A. (2013). Neonatal stress modulates sickness behavior: role for proinflammatory cytokines, *Journal of Neuroimmunology* 257(1–2), 59–66. doi: 10.1016/j.jneuroim.2013.02.009.

Barragan, A. A., Piñeiro, J. M., Schuenemann, G. M., Rajala-Schultz, P. J., Sanders, D. E., Lakritz, J. and Bas, S. (2018). Assessment of daily activity patterns and biomarkers of pain, inflammation, and stress in lactating dairy cows diagnosed with clinical metritis, *Journal of Dairy Science* 101(9), 8248–8258. doi: 10.3168/jds.2018-14510.

Beaver, A., Meagher, R. K., von Keyserlingk, M. A. G. and Weary, D. M. (2019). Invited review: a systematic review of the effects of early separation on dairy cow and calf health, *Journal of Dairy Science* 102(7), 5784–5810. doi: 10.3168/jds.2018-15603.

Bell, A. W. (1995). Regulation of organic nutrient metabolism during transition from late pregnancy to early lactation, *Journal of Animal Science* 73(9), 2804–2819. doi: 10.2527/1995.7392804x.

Blatteis, C. M. (2003). Fever: pathological or physiological, injurious or beneficial?, *Journal of Thermal Biology* 28(1), 1-13. doi: 10.1016/S0306-4565(02)00034-7.

Campler, M., Munksgaard, L., Jensen, M. B., Weary, D. M. and von Keyserlingk, M. A. (2014). Short communication: flooring preferences of dairy cows at calving, *Journal of Dairy Science* 97(2), 892-896. doi: 10.3168/jds.2013-7253.

Chagas, L. M., Bass, J. J., Blache, D., Burke, C. R., Kay, J. K., Lindsay, D. R., Lucy, M. C., Martin, G. B., Meier, S., Rhodes, F. M., Roche, J. R., Thatcher, W. W. and Webb, R. (2007). Invited review: new perspectives on the roles of nutrition and metabolic priorities in the subfertility of high-producing dairy cows, *Journal of Dairy Science* 90(9), 4022-4032. doi: 10.3168/jds.2006-852.

Chebel, R. C., Silva, P. R. B., Endres, M. I., Ballou, M. A. and Luchterhand, K. L. (2016). Social stressors and their effects on immunity and health of periparturient dairy cows, *Journal of Dairy Science* 99(4), 3217-3228. doi: 10.3168/jds.2015-10369.

Chebel, R. C., Mendonça, L. G. D. and Baruselli, P. S. (2018). Association between body condition score change during the dry period and postpartum health and performance, *Journal of Dairy Science* 101(5), 4595-4614. doi: 10.3168/jds.2017-13732.

Cook, N. B. and Nordlund, K. V. (2004). Managing the transition cow to optimize health and productivity, *Veterinary Clinics of North America: Food Animal Practice* 20(3), xi-xii. doi: 10.1016/j.cvfa.2004.06.007.

Cyples, J. A., Fitzpatrick, C. E., Leslie, K. E., DeVries, T. J., Haley, D. B. and Chapinal, N. (2012). Short communication: the effects of experimentally induced Escherichia coli clinical mastitis on lying behavior of dairy cows, *Journal of Dairy Science* 95(5), 2571-2575. doi: 10.3168/jds.2011-5135.

Dantzer, R. and Kelley, K. W. (2007). Twenty years of research on cytokine-induced sickness behavior, *Brain, Behavior, and Immunity* 21(2), 153-160. doi: 10.1016/j.bbi.2006.09.006.

Dechow, C. D. and Goodling, R. C. (2008). Mortality, culling by sixty days in milk, and production profiles in high- and low-survival Pennsylvania herds, *Journal of Dairy Science* 91(12), 4630-4639. doi: 10.3168/jds.2008-1337.

DeVries, T. J., von Keyserlingk, M. A. G. and Weary, D. M. (2004). Effect of feeding space on the inter-cow distance, aggression, and feeding behavior of free-stall housed lactating dairy cows, *Journal of Dairy Science* 87(5), 1432-1438. doi: 10.3168/jds.S0022-0302(04)73293-2.

De Vliegher, S., Fox, L. K., Piepers, S., McDougall, S. and Barkema, H. W. (2012). Invited review: mastitis in dairy heifers: nature of the disease, potential impact, prevention, and control, *Journal of Dairy Science* 95(3), 1025-1040. doi: 10.3168/jds.2010-4074.

Drackley, J. K. (1999). ADSA Foundation Scholar Award. Biology of dairy cows during the transition period: the final frontier?, *Journal of Dairy Science* 82(11), 2259-2273. doi: 10.3168/jds.S0022-0302(99)75474-3.

Drillich, M., Voigt, D., Forderung, D. and Heuwieser, W. (2007). Treatment of acute puerperal metritis with flunixin meglumine in addition to antibiotic treatment, *Journal of Dairy Science* 90(8), 3758-3763. doi: 10.3168/jds.2007-0052.

Eckelkamp, E. A. (2019). Invited review: current state of wearable precision dairy technologies in disease detection, *Applied Animal Science* 35(2), 209-220. doi: 10.15232/AAS.2018-01801.

Edwards, E. M., Krawczel, P. D., Dann, H. M., Schneider, L. G., Whitlock, B. and Proudfoot, K. L. (2020). Calving location preference and changes in lying and exploratory behaviour of preparturient dairy cattle with access to pasture, *Journal of Dairy Science* 103(6), 5455–5465. doi: 10.3168/jds.2019-17218.

Elbers, A. R. W., Miltenburg, J. D., De Lange, D., Crauwels, A. P., Barkema, H. W. and Schukken, Y. H. (1998). Risk factors for clinical mastitis in a random sample of dairy herds from the southern part of the Netherlands, *Journal of Dairy Science* 81(2), 420–426. doi: 10.3168/jds.S0022-0302(98)75592-4.

Endres, M. I. and Espejo, L. A. (2010). Feeding management and characteristics of rations for high-producing dairy cows in freestall herds, *Journal of Dairy Science* 93(2), 822–829. doi: 10.3168/jds.2008-2007.

Faulkner, P. M. and Weary, D. M. (2000). Reducing pain after dehorning in dairy calves, *Journal of Dairy Science* 83(9), 2037–2041. doi: 10.3168/jds.S0022-0302(00)75084-3.

Fitzpatrick, C. E., Chapinal, N., Petersson-Wolfe, C. S., DeVries, T. J., Kelton, D. F., Duffield, T. F. and Leslie, K. E. (2013). The effect of meloxicam on pain sensitivity, rumination time, and clinical signs in dairy cows with endotoxin-induced clinical mastitis, *Journal of Dairy Science* 96(5), 2847–2856. doi: 10.3168/jds.2012-5855.

Fogsgaard, K. K., Røntved, C. M., Sørensen, P. and Herskin, M. S. (2012). Sickness behavior in dairy cows during Escherichia coli mastitis, *Journal of Dairy Science* 95(2), 630–638. doi: 10.3168/jds.2011-4350.

Fogsgaard, K. K., Bennedsgaard, T. W. and Herskin, M. S. (2015). Behavioral changes in freestall-housed dairy cows with naturally occurring clinical mastitis, *Journal of Dairy Science* 98(3), 1730–1738. doi: 10.3168/jds.2014-8347.

Fraser, D., Weary, D. M., Pajor, E. A. and Milligan, B. N. (1997). A scientific conception of animal welfare that reflects ethical concerns, *Animal Welfare* 6(3), 187–205.

Fujiwara, M., Haskell, M. J., Macrae, A. I. and Rutherford, K. M. D. (2018). Survey of dry cow management on UK commercial dairy farms, *The Veterinary Record* 183(9), 297. doi: 10.1136/vr.104755.

Giuliodori, M. J., Magnasco, R. P., Becu-Villalobos, D., Lacau-Mengido, I. M., Risco, C. A. and de la Sota, R. L. (2013). Metritis in dairy cows: risk factors and reproductive performance, *Journal of Dairy Science* 96(6), 3621–3631. doi: 10.3168/jds.2012-5922.

Glaser, R., Kiecolt-Glaser, J. K., Bonneau, R. H., Malarkey, W., Kennedy, S. and Hughes, J. (1992). Stress-induced modulation of the immune response to recombinant hepatitis B vaccine, *Psychosomatic Medicine* 54(1), 22–29. Available at: http://journals.lww.com/psychosomaticmedicine/Fulltext/1992/01000/Stress_induced_modulation_of_the_immune_response.5.aspx.

Goff, J. P. (2008). The monitoring, prevention, and treatment of milk fever and subclinical hypocalcemia in dairy cows, *Veterinary Journal* 176(1), 50–57. doi: 10.1016/j.tvjl.2007.12.020.

González, M., Yabuta, A. K. and Galindo, F. (2003). Behaviour and adrenal activity of first parturition and multiparous cows under a competitive situation, *Applied Animal Behaviour Science* 83(4), 259–266. doi: 10.1016/S0168-1591(03)00037-6.

Grummer, R. R. (1995). Impact of changes in organic nutrient metabolism on feeding the transition dairy cow, *Journal of Animal Science* 73(9), 2820–2833. doi: 10.2527/1995.7392820x.

Grummer, R. R. (2008). Nutritional and management strategies for the prevention of fatty liver in dairy cattle, *Veterinary Journal* 176(1), 10–20. doi: 10.1016/j.tvjl.2007.12.033.

Grummer, R. R., Mashek, D. G. and Hayirli, A. (2004). Dry matter intake and energy balance in the transition period, *The Veterinary Clinics of North America. Food Animal Practice* 20(3), 447–470. doi: 10.1016/j.cvfa.2004.06.013.

Hammon, D. S., Evjen, I. M., Dhiman, T. R., Goff, J. P. and Walters, J. L. (2006). Neutrophil function and energy status in Holstein cows with uterine health disorders, *Veterinary Immunology and Immunopathology* 113(1–2), 21–29. doi: 10.1016/j.vetimm.2006.03.022.

Hart, B. L. (1988). Biological basis of the behavior of sick animals, *Neuroscience and Biobehavioral Reviews* 12(2), 123–137. doi: 10.1016/S0149-7634(88)80004-6.

Hennessy, M. B., Deak, T., Schiml-Webb, P. A., Carlisle, C. W. and O'Brien, E. (2010). Maternal separation produces, and a second separation enhances, core temperature and passive behavioral responses in guinea pig pups, *Physiology and Behavior* 100(4), 305–310. doi: 10.1016/j.physbeh.2010.02.024.

Hennessy, M. B., Deak, T. and Schiml, P. A. (2014). Sociality and sickness: have cytokines evolved to serve social functions beyond times of pathogen exposure?, *Brain, Behavior, and Immunity* 37, 15–20. doi: 10.1016/j.bbi.2013.10.021.

Herdt, T. H. (2000). Ruminant adaptation to negative energy balance: influences on the etiology of ketosis and fatty liver, *Veterinary Clinics of North America: Food Animal Practice* 16(2), 215–230. doi: 10.1016/S0749-0720(15)30102-X.

Humer, E., Petri, R. M., Aschenbach, J. R., Bradford, B. J., Penner, G. B., Tafaj, M., Südekum, K. H. and Zebeli, Q. (2018). Invited review: practical feeding management recommendations to mitigate the risk of subacute ruminal acidosis in dairy cattle, *Journal of Dairy Science* 101(2), 872–888. doi: 10.3168/jds.2017-13191.

Huzzey, J. M., von Keyserlingk, M. A. G. and Weary, D. M. (2005). Changes in feeding, drinking, and standing behavior of dairy cows during the transition period, *Journal of Dairy Science* 88(7), 2454–2461. doi: 10.3168/jds.S0022-0302(05)72923-4.

Huzzey, J. M., DeVries, T. J., Valois, P. and von Keyserlingk, M. A. (2006). Stocking density and feed barrier design affect the feeding and social behavior of dairy cattle, *Journal of Dairy Science* 89(1), 126–133. doi: 10.3168/jds.S0022-0302(06)72075-6.

Huzzey, J. M., Veira, D. M., Weary, D. M. and von Keyserlingk, M. A. (2007). Prepartum behavior and dry matter intake identify dairy cows at risk for metritis, *Journal of Dairy Science* 90(7), 3220–3233. doi: 10.3168/jds.2006-807.

Huzzey, J. M., Nydam, D. V., Grant, R. J. and Overton, T. R. (2012a). The effects of overstocking Holstein dairy cattle during the dry period on cortisol secretion and energy metabolism, *Journal of Dairy Science* 95(8), 4421–4433. doi: 10.3168/jds.2011-5037.

Huzzey, J. M., Grant, R. J. and Overton, T. R. (2012b). Short communication: relationship between competitive success during displacements at an overstocked feed bunk and measures of physiology and behavior in Holstein dairy cattle, *Journal of Dairy Science* 95(8), 4434–4441. doi: 10.3168/jds.2011-5038.

Ingvartsen, K. L., Dewhurst, R. J. and Friggens, N. C. (2003). On the relationship between lactational performance and health: is it yield or metabolic imbalance that cause production diseases in dairy cattle? A position paper, *Livestock Production Science* 83(2–3), 277–308. doi: 10.1016/S0301-6226(03)00110-6.

Jensen, M. B. (2012). Behaviour around the time of calving in dairy cows, *Applied Animal Behaviour Science* 139(3–4), 195–202. doi: 10.1016/j.applanim.2012.04.002.

Jensen, M. B. and Proudfoot, K. L. (2017). Effect of group size and health status on behavior and feed intake of multiparous dairy cows in early lactation, *Journal of Dairy Science* 100(12), 9759-9768. doi: 10.3168/jds.2017-13035.

Jensen, M. B. and Rørvang, M. V. (2018). The degree of visual cover and location of birth fluids affect dairy cows' choice of calving site, *Journal of Dairy Science* 101(10), 9483–9492. doi: 10.3168/jds.2018-14724.

King, M. T. M., LeBlanc, S. J., Pajor, E. A., Wright, T. C. and DeVries, T. J. (2018). Behavior and productivity of cows milked in automated systems before diagnosis of health disorders in early lactation, *Journal of Dairy Science* 101(5), 4343–4356. doi: 10.3168/JDS.2017-13686.

Lassen, J., Sandøe, P. and Forkman, B. (2006). Happy pigs are dirty! - conflicting perspectives on animal welfare, *Livestock Science* 103(3), 221–230. doi: 10.1016/j.livsci.2006.05.008.

LeBlanc, S. (2010). Monitoring metabolic health of dairy cattle in the transition period, *Journal of Reproduction and Development* 56(Suppl.), S29–S35. doi: 10.1262/jrd.1056S29.

LeBlanc, S. J., Lissemore, K. D., Kelton, D. F., Duffield, T. F. and Leslie, K. E. (2006). Major advances in disease prevention in dairy cattle, *Journal of Dairy Science* 89(4), 1267–1279. doi: 10.3168/jds.S0022-0302(06)72195-6.

Lidfors, L. M., Moran, D., Jung, J., Jensen, P. and Castren, H. (1994). Behaviour at calving and choice of calving place in cattle kept in different environments, *Applied Animal Behaviour Science* 42(1), 11–28. doi: 10.1016/0168-1591(94)90003-5.

Lobeck-Luchterhand, K. M., Silva, P. R., Chebel, R. C. and Endres, M. I. (2014). Effect of prepartum grouping strategy on displacements from the feed bunk and feeding behavior of dairy cows, *Journal of Dairy Science* 97(5), 2800–2807. doi: 10.3168/jds.2013-7401.

Lobeck-Luchterhand, K. M., Silva, P. R., Chebel, R. C. and Endres, M. I. (2015). Effect of stocking density on social, feeding, and lying behavior of prepartum dairy animals, *Journal of Dairy Science* 98(1), 240–249. doi: 10.3168/jds.2014-8492.

Mainau, E. and Manteca, X. (2011). Pain and discomfort caused by parturition in cows and sows, *Applied Animal Behaviour Science* 135(3), 241–251. doi: 10.1016/j.applanim.2011.10.020.

Meagher, R. K., Beaver, A., Weary, D. M. and von Keyserlingk, M. A. G. (2019). Invited review: a systematic review of the effects of prolonged cow–calf contact on behavior, welfare, and productivity, *Journal of Dairy Science* 102(7), 5765–5783. doi: 10.3168/jds.2018-16021.

Medrano-Galarza, C., Gibbons, J., Wagner, S., de Passillé, A. M. and Rushen, J. (2012). Behavioral changes in dairy cows with mastitis, *Journal of Dairy Science* 95(12), 6994–7002. doi: 10.3168/jds.2011-5247.

Mee, J. F. (2008). Prevalence and risk factors for dystocia in dairy cattle: a review, *Veterinary Journal* 176(1), 93–101. doi: 10.1016/j.tvjl.2007.12.032.

Miedema, H. M., Cockram, M. S., Dwyer, C. M. and Macrae, A. I. (2011a). Behavioural predictors of the start of normal and dystocic calving in dairy cows and heifers, *Applied Animal Behaviour Science* 132(1–2), 14–19. doi: 10.1016/j.applanim.2011.03.003.

Miedema, H. M., Cockram, M. S., Dwyer, C. M. and Macrae, A. I. (2011b). Changes in the behaviour of dairy cows during the 24h before normal calving compared with behaviour during late pregnancy, *Applied Animal Behaviour Science* 131(1–2), 8–14. doi: 10.1016/j.applanim.2011.01.012.

Miltenburg, C. L., Duffield, T. F., Bienzle, D., Scholtz, E. L. and LeBlanc, S. J. (2018). The effect of prepartum feeding and lying space on metabolic health and immune function, *Journal of Dairy Science* 101(6), 5294–5306. doi: 10.3168/jds.2017-13481.

Mulligan, F. J. and Doherty, M. L. (2008). Production diseases of the transition cow, *Veterinary Journal* 176(1), 3–9. doi: 10.1016/j.tvjl.2007.12.018.

Oltenacu, P. A. and Broom, D. M. (2010). The impact of genetic selection for increased milk yield on the welfare of dairy cows, *Animal Welfare* 19(S), 39–49.

Ospina, P. A., Nydam, D. V., Stokol, T. and Overton, T. R. (2010). Evaluation of nonesterified fatty acids and beta-hydroxybutyrate in transition dairy cattle in the northeastern United States: critical thresholds for prediction of clinical diseases, *Journal of Dairy Science* 93(2), 546–554. doi: 10.3168/jds.2009-2277.

Overton, T. R. and Waldron, M. R. (2004). Nutritional management of transition dairy cows: strategies to optimize metabolic health, *Journal of Dairy Science* 87(E. Suppl.), E105–E119. doi: 10.3168/jds.S0022-0302(04)70066-1.

Peeler, E. J., Otte, M. J. and Esslemont, R. J. (1994). Inter-relationships of periparturient diseases in dairy cows, *The Veterinary Record* 134(6), 129–132. doi: 10.1136/vr.134.6.129.

Pohl, A., Bertulat, S., Borchardt, S., Burfeind, O. and Heuwieser, W. (2016). Randomized, controlled clinical trial on the efficacy of nonsteroidal antiinflammatory drugs for the treatment of acute puerperal metritis in dairy cows, *Journal of Dairy Science* 99(10), 8241–8249. doi: 10.3168/jds.2015-10775.

Proudfoot, K. and Habing, G. (2015). Social stress as a cause of diseases in farm animals: current knowledge and future directions, *The Veterinary Journal* 206(1), 15–21. doi: 10.1016/j.tvjl.2015.05.024.

Proudfoot, K. L., Veira, D. M., Weary, D. M. and von Keyserlingk, M. A. (2009a). Competition at the feed bunk changes the feeding, standing, and social behavior of transition dairy cows, *Journal of Dairy Science* 92(7), 3116–3123. doi: 10.3168/jds.2008-1718.

Proudfoot, K. L., Huzzey, J. M. and von Keyserlingk, M. A. G. (2009b). The effect of dystocia on the dry matter intake and behavior of Holstein cows, *Journal of Dairy Science* 92(10), 4937–4944. doi: 10.3168/jds.2009-2135.

Proudfoot, K. L., Jensen, M. B., Heegaard, P. M. and von Keyserlingk, M. A. (2013). Effect of moving dairy cows at different stages of labor on behavior during parturition, *Journal of Dairy Science* 96(3), 1638–1646. doi: 10.3168/jds.2012-6000.

Proudfoot, K. L., Weary, D. M. and von Keyserlingk, M. A. (2014a). Maternal isolation behavior of Holstein dairy cows kept indoors, *Journal of Animal Science* 92(1), 277–281. doi: 10.2527/jas.2013-6648.

Proudfoot, K. L., Jensen, M. B., Weary, D. M. and von Keyserlingk, M. A. (2014b). Dairy cows seek isolation at calving and when ill, *Journal of Dairy Science* 97(5), 2731–2739. doi: 10.3168/jds.2013-7274.

Proudfoot, K. L., Weary, D. M., LeBlanc, S. J., Mamedova, L. K. and von Keyserlingk, M. A. G. (2018). Exposure to an unpredictable and competitive social environment affects behavior and health of transition dairy cows, *Journal of Dairy Science* 101(10), 9309–9320. doi: 10.3168/jds.2017-14115.

Reinhardt, T. A., Lippolis, J. D., McCluskey, B. J., Goff, J. P. and Horst, R. L. (2011). Prevalence of subclinical hypocalcemia in dairy herds, *Veterinary Journal* 188(1), 122–124. doi: 10.1016/j.tvjl.2010.03.025.

Remnant, J. G., Tremlett, A., Huxley, J. N. and Hudson, C. D. (2017). Clinician attitudes to pain and use of analgesia in cattle: where are we 10 years on? *Veterinary Record*. doi: 10.1136/vr.104428.

Rørvang, M. V., Herskin, M. S. and Jensen, M. B. (2017). Dairy cows with prolonged calving seek additional isolation, *Journal of Dairy Science* 100(4), 2967–2975. doi: 10.3168/jds.2016-11989.

Rørvang, M. V., Herskin, M. S. and Jensen, M. B. (2018a). The motivation-based calving facility: social and cognitive factors influence isolation seeking behaviour of Holstein dairy cows at calving, *PLoS ONE* 13(1), e0191128. doi: 10.1371/journal.pone.0191128.

Rørvang, M. V., Nielsen, B. L., Herskin, M. S. and Jensen, M. B. (2018b). Prepartum maternal behavior of domesticated cattle: a comparison with managed, feral, and wild ungulates, *Frontiers in Veterinary Science* 5, 45. doi: 10.3389/fvets.2018.00045.

Ruprechter, G., Adrien, M. L., Larriestra, A., Meotti, O., Batista, C., Meikle, A. and Noro, M. (2018). Metabolic predictors of peri-partum diseases and their association with parity in dairy cows, *Research in Veterinary Science*. W. B. Saunders 118, 191–198. doi: 10.1016/j.rvsc.2018.02.005.

Schirmann, K., Chapinal, N., Weary, D. M., Heuwieser, W. and von Keyserlingk, M. A. (2011). Short-term effects of regrouping on behavior of prepartum dairy cows, *Journal of Dairy Science* 94(5), 2312–2319. doi: 10.3168/jds.2010-3639.

Schirmann, K., Weary, D. M., Heuwieser, W., Chapinal, N., Cerri, R. L. A. and von Keyserlingk, M. A. G. (2016). Short communication: rumination and feeding behaviors differ between healthy and sick dairy cows during the transition period, *Journal of Dairy Science* 99(12), 9917–9924. doi: 10.3168/jds.2015-10548.

Segerstrom, S. C. and Miller, G. E. (2004). Psychological stress and the human immune system: a meta-analytic study of 30 years of inquiry, *Psychological Bulletin* 130(4), 601–630. doi: 10.1037/0033-2909.130.4.601.

Sepúlveda-Varas, P., Huzzey, J. M., Weary, D. M. and von Keyserlingk, M. A. G. (2013). Behaviour, illness and management during the periparturient period in dairy cows, *Animal Production Science* 53(9), 988–999. doi: 10.1071/AN12286.

Sepúlveda-Varas, P., Weary, D. M. and von Keyserlingk, M. A. G. (2014). Lying behavior and postpartum health status in grazing dairy cows, *Journal of Dairy Science* 97(10), 6334–6343. doi: 10.3168/jds.2014-8357.

Shakhar, K. and Shakhar, G. (2015). Why do we feel sick when infected–can altruism play a role?, *PLoS Biology*. Public Library of Science 13(10), e1002276. doi: 10.1371/journal.pbio.1002276.

Shin, D. H., Jeong, J. K., Choi, I. S., Moon, S. H., Lee, S. C., Kang, H. G., Park, S. B. and Kim, I. H. (2018). Associations between serum haptoglobin concentration and peri- and postpartum disorders, milk yield, and reproductive performance in dairy cows, *Livestock Science* 213, 14–18. doi: 10.1016/j.livsci.2018.04.015.

Siivonen, J., Taponen, S., Hovinen, M., Pastell, M., Lensink, B. J., Pyörälä, S. and Hänninen, L. (2011). Impact of acute clinical mastitis on cow behaviour, *Applied Animal Behaviour Science* 132(3–4), 101–106. doi: 10.1016/j.applanim.2011.04.005.

Silva, P. R. B., Moraes, J. G. N., Mendonça, L. G. D., Scanavez, A. A., Nakagawa, G., Ballou, M. A., Walcheck, B., Haines, D., Endres, M. I. and Chebel, R. C. (2013a). Effects of weekly regrouping of prepartum dairy cows on innate immune response and antibody concentration, *Journal of Dairy Science* 96(12), 7649–7657. doi: 10.3168/jds.2013-6752.

Silva, P. R. B., Moraes, J. G. N., Mendonça, L. G. D., Scanavez, A. A., Nakagawa, G., Fetrow, J., Endres, M. I. and Chebel, R. C. (2013b). Effects of weekly regrouping of prepartum dairy cows on metabolic, health, reproductive, and productive parameters, *Journal of Dairy Science* 96(7), 4436–4446. doi: 10.3168/jds.2012-6464.

Silva, P. R. B., Lobeck-Luchterhand, K. M., Cerri, R. L., Haines, D. M., Ballou, M. A., Endres, M. I. and Chebel, R. C. (2016). Effects of prepartum stocking density on innate and adaptive leukocyte responses and serum and hair cortisol concentrations, *Veterinary Immunology and Immunopathology* 169, 39–46. doi: 10.1016/j.vetimm.2015.11.007.

Sordillo, L. M. (2016). Nutritional strategies to optimize dairy cattle immunity, *Journal of Dairy Science* 99(6), 4967–4982. doi: 10.3168/JDS.2015-10354.

Sova, A. D., LeBlanc, S. J., McBride, B. W. and DeVries, T. J. (2013). Associations between herd-level feeding management practices, feed sorting, and milk production in freestall dairy farms, *Journal of Dairy Science* 96(7), 4759–4770. doi: 10.3168/jds.2013-6679.

Stojkov, J., von Keyserlingk, M. A., Marchant-Forde, J. N. and Weary, D. M. (2015). Assessment of visceral pain associated with metritis in dairy cows, *Journal of Dairy Science* 98(8), 5352–5361. doi: 10.3168/jds.2014-9296.

Suthar, V., Burfeind, O., Bonk, S., Voigtsberger, R., Keane, C. and Heuwieser, W. (2012). Factors associated with body temperature of healthy Holstein dairy cows during the first 10 days in milk, *Journal of Dairy Research* 79(2), 135–142. doi: 10.1017/S0022029911000896.

Suthar, V. S., Canelas-Raposo, J., Deniz, A. and Heuwieser, W. (2013). Prevalence of subclinical ketosis and relationships with postpartum diseases in European dairy cows, *Journal of Dairy Science* 96(5), 2925–2938. doi: 10.3168/jds.2012-6035.

Tatone, E. H., Duffield, T. F., LeBlanc, S. J., DeVries, T. J. and Gordon, J. L. (2017). Investigating the within-herd prevalence and risk factors for ketosis in dairy cattle in Ontario as diagnosed by the test-day concentration of β-hydroxybutyrate in milk, *Journal of Dairy Science* 100(2), 1308–1318. doi: 10.3168/jds.2016-11453.

Te Velde, H., Aarts, N. and Van Woerkum, C. (2002). Dealing with ambivalence: farmers' and consumers' perceptions of animal welfare in livestock breeding, *Journal of Agricultural and Environmental Ethics* 15(2), 203–219. doi: 10.1023/A:1015012403331.

Thomsen, P. T., Anneberg, I. and Herskin, M. S. (2012). Differences in attitudes of farmers and veterinarians towards pain in dairy cows, *The Veterinary Journal* 194(1), 94–97. doi: 10.1016/j.tvjl.2012.02.025.

Urton, G., von Keyserlingk, M. A. G. and Weary, D. M. (2005). Feeding behavior identifies dairy cows at risk for metritis, *Journal of Dairy Science* 88(8), 2843–2849. doi: 10.3168/jds.S0022-0302(05)72965-9.

Verbeke, W. (2009). Stakeholder, citizen and consumer interests in farm animal welfare, *Animal Welfare* 18(4), 325–333. doi: 10.2752/089279306785415439.

Vergara, C. F., Döpfer, D., Cook, N. B., Nordlund, K. V., McArt, J. A., Nydam, D. V. and Oetzel, G. R. (2014). Risk factors for postpartum problems in dairy cows: explanatory and predictive modeling, *Journal of Dairy Science* 97(7), 4127–4140. doi: 10.3168/jds.2012-6440.

von Keyserlingk, M. A. G., Rushen, J., de Passillé, A. M. and Weary, D. M. (2009). Invited review: the welfare of dairy cattle--key concepts and the role of science, *Journal of Dairy Science* 92(9), 4101–4111. doi: 10.3168/jds.2009-2326.

Weary, D. M., Niel, L., Flower, F. C. and Fraser, D. (2006). Identifying and preventing pain in animals, *Applied Animal Behaviour Science* 100(1-2), 64–76. doi: 10.1016/j.applanim.2006.04.013.

Weary, D. M., Huzzey, J. M. and von Keyserlingk, M. A. G. (2009). Board-invited review: using behavior to predict and identify ill health in animals, *Journal of Animal Science* 87(2), 770–777. doi: 10.2527/jas.2008-1297.

Wenz, J. R., Moore, D. A. and Kasimanickam, R. (2011). Factors associated with the rectal temperature of Holstein dairy cows during the first 10 days in milk, *Journal of Dairy Science* 94(4), 1864–1872. doi: 10.3168/jds.2010-3924.

Zebeli, Q., Ghareeb, K., Humer, E., Metzler-Zebeli, B. U. and Besenfelder, U. (2015). Nutrition, rumen health and inflammation in the transition period and their role on overall health and fertility in dairy cows, *Research in Veterinary Science* 103, 126–136. doi: 10.1016/j.rvsc.2015.09.020.

Zerbe, H., Schneider, N., Leibold, W., Wensing, T., Kruip, T. A. and Schuberth, H. J. (2000). Altered functional and immunophenotypical properties of neutrophilic granulocytes in postpartum cows associated with fatty liver, *Theriogenology* 54(5), 771–786. doi: 10.1016/S0093-691X(00)00389-7.

Chapter 10

Optimizing welfare in transport and slaughter of cattle

Jan Shearer, Iowa State University, USA

1 Introduction

Safeguarding the welfare of animals during transport and at slaughter requires an understanding of the risks encountered during each phase of the journey and at the destination. Pre-transport assessment of an animal's fitness for travel can reduce problems during transport and at slaughter. If the trip or time in transit is likely to be lengthy, it is important that animals are well fed and hydrated prior to departure. During transport, the welfare of cattle is affected by the experience of the driver, time and distance travelled and weather conditions. The scarcity of slaughter facilities throughout the United States has resulted in longer travel times and distances, contributing to driver fatigue and a greater potential for livestock hauler accidents. The welfare of cattle at slaughter is affected by the condition of animals on arrival and opportunities for them to rehydrate and rest in lairage. Proper scheduling should prevent the need for animals to remain in lairage for extended periods prior to slaughter.

http://dx.doi.org/10.19103/AS.2020.0084.13

Cattle handling, movement of animals to the stunning box, stunning accuracy and slaughter procedures, also affect the welfare of cattle at abattoirs.

To provide context for the development of strategies to optimize the welfare of cattle during the pre-transport, transport and slaughter phases, it is important to understand why animals are removed from herds; what are the common modes of transport; and, what are some of the current observations regarding the condition of animals at slaughter? Not all cattle are culled for reasons of disease or injury. Most animals culled from herds are healthy but must be removed because of inferior performance relative to an available herd replacement. Dairy cows leave herds on average at a little less than 5 years of age, whereas the longevity of beef cows is approximately twice that of dairy cows at 10 or more years.

Modes of transport for healthy cattle culled from farms and ranches to markets and slaughter are generally via semi-pulled potbelly and light truck gooseneck or bumper-pull type trailers. Compromised cattle should be hauled to the nearest possible slaughter facility and in trailers where they can be isolated from other animals. Muscling and body condition of cull dairy cows arriving at slaughter plants is consistently low. This not only reduces carcass value but increases susceptibility to bruising during transport.

Although the incidence of obvious physical defects in cattle at slaughter is low, more than one-third of hides are damaged by the practice of branding. Hide-off evaluation of carcasses at slaughter reveals that roughly two-thirds of cattle carcasses require some degree of trimming because of bruising. The incidence of 'dark cutters' (i.e. carcasses that are dark in color) is associated with pre-slaughter stress. Branding, bruising and dark cutting beef are important indicators of poor welfare. They are not only very costly problems, but they are also largely preventable.

2 Culling and permanent herd removals

2.1 Voluntary versus involuntary culling

Culling is an animal removal strategy intended to improve productivity and profitability in dairy and beef cattle herds. Reasons for culling are often described in terms of 'voluntary' or 'involuntary'. Cows that are removed from herds for voluntary reasons are those removed at the will of the owner. Involuntary reasons for removing an animal from the herd are those against the will or conscious control of the owner. Dairy cows that leave herds for voluntary reasons are primarily influenced by milk price, market or slaughter value of cows, cost and availability of replacements and housing capacity. Involuntary herd removals are generally for health-related conditions that reduce an animal's performance below cost-benefit breakeven levels. Depending upon the specific condition, these animals present the greatest welfare concern in the context of transport to markets and slaughter.

2.2 Surplus replacements and bull sales

In situations where the dairy has an over-supply of replacements, it may elect to sell animals for 'dairy purposes', that is, sell animals to other dairies for making milk. Some dairies raise bulls for sale to herds that rely on natural service for some if not all of their cattle breeding needs. Purebred beef herds with an excess of replacements may sell heifers, cows or bulls to other herds seeking genetic improvement. Sales such as these do not fit the cull animal criterion. These animals are not removed for reasons of poor performance; rather, they leave because the herd has animals in excess of its needs or is in the business of marketing superior genetics.

2.3 Culling based upon performance

The primary product of dairy farming is milk; in the cow/calf industry, the product is a calf on an annual basis. Calves from cow/calf operations may go directly to a feedlot, whereas others may make an intermediate stop at a backgrounding operation, where they are readied for movement to the final production phase in a feedlot. The feedlot's objective is to produce a finished steer or heifer ready for slaughter.

Performance and production efficiency are key factors in the profitability of each enterprise. A dairy cow that fails to produce milk at a level above the cost of maintaining her becomes a cull and is moved to slaughter. Similar decisions are made for a beef cow that fails to become pregnant or calves in the stocker or feedlot industries that fail to grow or gain at profitable levels. Bulls are valuable as long as they are fertile and physically capable of producing pregnancies in cows. When their capacity for producing pregnancies in cows is lost, they too are moved onto slaughter. This is the truest sense of the term 'cull', that is, to remove an animal from a herd that is less productive so that a more profitable animal may replace it.

2.4 Removal (i.e. cull) rates for dairy and beef herds

The annual herd removal rate for dairy cattle operations in the United States is approximately 30% (Geiger, 2016). Rates vary considerably and are influenced by multiple factors including parity, milk production, fertility and cow health. Some herds may temporarily reduce or adjust cull rates when planning for herd expansion. Another driver of cull rates is the availability of herd replacements. The advent of sexed semen has created a surplus of dairy heifer replacements allowing for more stringent culling of older, less profitable cows. De Vries calculated that the normal productive life of a dairy cow in the United States was 2.63 years or 31.6 months. Assuming an average age at first calving of 25.5 months, plus a cow's average productive life of 31.6 months, the average dairy cow remains in the herd for approximately 57.1 months or 4.8 years of age before being removed from the herd (De Vries, 2013).

Annual cull rates for cow/calf herds are also variable, but estimated to be in the range of 10–20%. A beef cow must successfully deliver and raise a calf each year, and in addition be healthy and fertile enough to become pregnant. Failure to accomplish these feats makes her a candidate for culling from the herd. Over half of the cows (52.6%) sold from herds in 2007 were 10 years old or older than that. According to USDA statistics, 41.8% of beef cattle operations reported culling cows that did not become pregnant. Other reasons for culling cows included: age and bad teeth 55.7%, temperament 16.6%, other (unspecified) reproductive health problem 13.4%, economics (drought, need to reduce herd size, and market conditions) 10.9%, producing calves 10.7%, physical unsoundness 9.6%, udder problems 9.2% and eyesight problems 7.1% (USDA, 2007–08, https://www.aphis.usda.gov/animal_health/nahms/beefcowcalf/downloads/beef0708/Beef0708_dr_PartIV_1.pdf). The removal of less-productive cows creates opportunities for herds with an ample number of healthy productive replacements.

3 Transport of cattle to markets and slaughter facilities

A rough estimate of the number of cattle transported to dairies, growing operations, livestock markets and/or slaughter plants likely exceeds 400 000 each day. The overwhelming majority are hauled in semis pulling potbelly trailers and pick-ups with a gooseneck or bumper-pull trailer in tow. In cattle states throughout the United States, it is virtually impossible to avoid encountering a livestock hauler transporting cattle or other livestock to some destination whether it be a market, packing plant or just from one farm to another.

3.1 Potbelly trailers

Feeder calves, market and slaughter cattle are often hauled in semi-pulled potbelly or modified potbelly trailers. The term potbelly refers to the 'drop down belly' design of the trailer that when loaded increases stability (reducing the potential for rollover accidents) and permits better space utilization. These trailers have an upper deck and cut gates that permit proper distribution of weight and cattle into the upper (top) and lower (bottom) nose, upper deck, belly, doghouse (also known as the jail), which is the upper compartment in the back of the trailer. Depending upon the number of axles some of these trailers can haul nearly 90 000 lb (40 823 kg).

Potbelly trailers can be modified into a fat cattle/feeder cattle combo trailer with the versatility to haul both feeder calves and finished beef cattle (Fig. 1). Fat trailers are designed to haul finished feedlot steers and heifers only (Fig. 2). The differences between these types of trailers include the presence

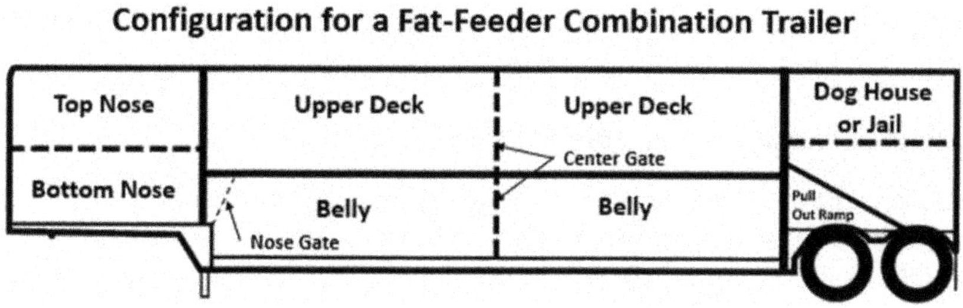

Figure 1 Configuration for a trailer designed to haul fat cattle and/or feeder steers and heifers (adapted from Master Cattle Transporter Guide, https://www.bqa.org/Media/BQA/Docs/master_cattle_transporter_guide-digital.pdf).

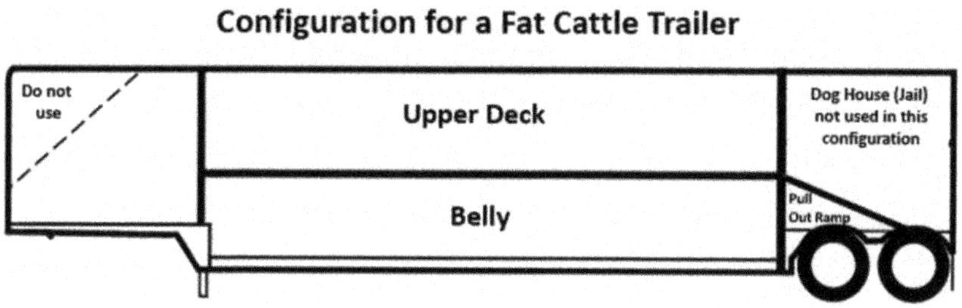

Figure 2 Configuration for a trailer designed to haul fat cattle only (adapted from Master Cattle Transporter Guide, https://www.bqa.org/Media/BQA/Docs/master_cattle_transporter_guide-digital.pdf).

or absence of the 'jail' or 'doghouse' in the upper rear compartment of the trailer. This compartment (the doghouse) is present in the fat/feeder combo configuration. Both configurations have a slide-in or fold-up ramp that leads to the upper deck above the belly. One feature of note is that the clearance height of the entrance into the 'belly', or lower compartment of the trailer is approximately 2–3 inches (5–8 cm) shorter in the fat/feeder combination trailers. This has important implications for taller cattle breeds such as Holsteins, which are more likely to injure themselves on entry and exit of from the belly of the trailer (Master Cattle Transporter Guide, National Beef Quality Assurance Guide for Cattle Transporters, https://www.bqa.org/Media/BQA/Docs/master_cattle_transporter_guide-digital.pdf).

In order to assess the number of traumatic events cattle may encounter in the process of unloading, researchers counted the number of times cattle hit any part of the trailer during the off-loading process at packing plants. A total of 9860 animals in 75 lots were observed at three slaughter plants being off-loaded from 275 trailers. The prevalence of traumatic events was highest for Holstein cattle hauled in fat/feeder combination trailers. Researchers also found a significantly higher prevalence of carcass bruising in Holstein cattle (76.6%) compared to 66.6% in beef breed cattle (Lee et al., 2017). The most likely reasons are the larger frame and height of Holsteins, which would naturally make them more susceptible to trauma and bruising.

3.2 Gooseneck and bumper-pull trailers

Gooseneck trailers have a long neck that reaches over the tailgate of the truck box to fit over a ball hitch in the truck bed. Livestock trailers are available in lengths ranging from 16 ft (4.9 m) to 36 ft (11 m). Trailer floors are either wood or in some cases aluminum. Depending upon the length of the trailer most will have at least one cut gate. An 8 ft (2.4 m) × 24 ft (7.3 m) trailer should accommodate 3–5 1200–1400 lb (544-635 kg) animals per compartment. Approximately 20–25% of the weight of the load should be on the hitch. This adds stability when the trailer is loaded because the weight is better distributed over the axles and truck chassis (Livestock trailer safety, 2012, http://www.extension.org/pages/64391/livestock-trailer-safety).

Bumper pull trailers attach to a ball hitch on the back bumper of the truck, which is connected to the truck's frame. Weight distribution is more critical with these trailers. Generally, 60% of the weight should be in front of the trailer axle(s) with around 10–15% of the total weight on the hitch. Insufficient weight on the hitch or tongue relative to the rear of the trailer can cause the trailer to swing the rear of the truck from side to side making it extremely difficult to control the vehicle (Livestock trailer safety, http://www.extension.org/pages/64391/livestock-trailer-safety) .

Truck and trailer maintenance (proper tire inflation and wheel bearing maintenance) is extremely important. Breakdowns with cattle onboard in hot and humid or extremely cold weather can be very detrimental (if not deadly) for the welfare of cattle on board. For trailers with wood floors, it is important to remember that wood floors eventually rot. If an animal happens to place a foot on a weak spot in the floor and it breaks through, it can have disastrous consequences (Stalsberg et al., https://fyi.extension.wisc.edu/wbic/files/2018/03/3T-handout-1done.pdf).

3.3 Transporting compromised cattle

Cull cows compromised by lameness or other condition, yet fit for human consumption and short-distance travel, should be transported to the nearest possible slaughter plant. They should not be transported to markets where there is the potential for long distance travel on the next leg of their journey. If several animals are being transported, the animals with the best mobility should be loaded first and toward the front of the trailer. Lame or weaker animals should be loaded last and kept in a separate section from other animals. This helps avoid further injury that may occur in transit from bullying by other animals or trampling should the animal fall.

Cull cows sold at markets are often hauled on potbelly trailers to their final destination at a packing plant. The hauler must carefully evaluate the size and fitness of these animals, particularly when long-distance travel may be required. Weak or lame animals may have difficulty walking up and down ramps upon entry and exit from trailers. In potbelly trailers, cattle must be able to walk up and down ramps into the belly of the trailer as well as to the upper deck. They must also adjust to co-mingling with other animals and vehicular movement during transport. For many reasons, the welfare of cull cows and particularly those with special needs requires additional care and handling (Cockram, 2019).

4 Trends in transport of dairy and beef cattle

NAHMS data for all operations studied found that 92% of dairies sold cows through livestock markets, auctions or stockyards. Thirty-seven percent of dairy operations sent cull cows directly to slaughter. Smaller operations tended to send cull cows to markets rather than direct to slaughter plants. As herd size increased, the average number of cows shipped to a market, auction or stockyard, or directly to a packer increased (USDA, 2018).

Cattle leaving herds by the number of lactations was 22.2% for first, 51.9% for second to fourth, and 25.9% for fifth or higher lactations. Of cows permanently removed, 9.5% were sold as replacements and 21.1% were sold for low milk production. Cows that left herds for involuntary reasons included 21.2% for infertility, 16.5% for clinical mastitis and 7.2% for lameness. By stage of lactation, researchers observed that approximately 20% of cows left herds in early lactation, 24% at mid-lactation and nearly 50% (49.3%) of cows were removed in late lactation. Cows with serious health conditions and low milk production generally left herds early in lactation (USDA, 2018, https://www.aphis.usda.gov/animal_health/nahms/dairy/downloads/dairy14/Dairy14_dr_PartIII.pdf).

In terms of distance travelled to a market, auction or stockyard, 14.3% of shipments were within 1 (1.6 km) to 9 (14.5 km) miles and 63.4% of shipments were within 10 (16 km) to 49 (79 km) miles. This means that (14.3 % + 63.4%) roughly, 80% of cattle travel less than 50 (80.5 km) miles to a market, which sounds reasonable until one considers that this is likely only the first stop on their journey. From here, animals will travel to a slaughter plant, which for cattle in certain parts of the West or Southeast could mean a very long trip on the second leg of their journey (Edwards-Callaway et al., 2019). NAHMS data indicates that 30% of dairies send cows to slaughter plants across state lines (USDA, 2018, https://www.aphis.usda.gov/animal_health/nahms/dairy/downloads/dairy14/Dairy14_dr_PartIII.pdf). However, depending upon the needs of the cattle buyer, cattle may be transported much further (Edwards-Callaway et al., 2019).

For shipments going directly to slaughter, 7.7% travelled from 1 (1.6 km) to 9 (14.5 km) miles, 42.8% from 10 (16 km) to 49 miles (79 km), 38.3% from 50 (80.5 km) to 249 (400 km) miles and slightly over 11.2% travelled more than 250 (402 km) miles. By simple calculation, approximately 50% of animals travel less than 50 miles (80.5 km), but more than 11% travel in excess of 250 miles (402 km). Any animal that exhibits weakness or immobility, yet can still be safely processed for human use, should be taken to the closest possible slaughter plant (USDA, 2018, https://www.aphis.usda.gov/animal_health/nahms/dairy/downloads/dairy14/Dairy14_dr_PartIII.pdf).

An excellent review of the current challenges to the welfare of cattle during transport to markets and slaughter in the United States was published by Edwards-Callaway (Edwards-Callaway et al., 2019).

4.1 Transport of beef cattle

According to the recent National Beef Quality Audit, cattle were found to be in transit for a mean of 6.7 h and travelled a mean distance of 283 miles (455.7 km). Three loads travelled more than 24 h and travelled distances greater than 994 miles (1600 km) (Harris et al., 2017). Long-distance travel is associated with an increased amount of carcass bruising, shrink (dehydration) and stress. Longer trips and transit times (i.e. beyond 24 h) also require stops in route to provide feed and water in accordance with the Beef Quality Assurance program guidelines. The maximum transportation times collected were 12 h for fed cattle and 39 h for cows/bulls. The longer and farther animals are transported the greater the impact on carcass quality traits (bruising and dark cutting beef) at slaughter.

The average number of cattle per load was 26, although load size ranged from 1 to 47 animals per trailer (Harris et al., 2017). Recommended minimum space requirements for cattle weighing 360 kg (800 lb), 454 kg (1000 lb), 545 kg (1200 lb) and 635 kg (1400 lb) are .97 m^2 (10.4 ft^2), 1.11 m^2 (12 ft^2), 1.35 m^2 (14.5 ft^2) and 1.67 m^2 (18 ft^2), respectively. Range in space allowance for cattle with horns is an additional .05 m^2 to .09 m^2 (0.5–1 ft^2) (Grandin, 2013). The average space allotment per animal was 2.3 m^2 (24.8 ft^2) for all loads studied indicating that recommendations for space per animal were met in the loads observed (Harris et al., 2017). Additional guidelines on trailer compartment dimensions and weight of the cattle hauled may be found in the Master Cattle Transporter Guide at: https://www.bqa.org/Media/BQA/Docs/master_cattle_transporter_guide-digital.pdf.

Trailer dimensions varied, as there were several types of trailers used. Pot belly trailers were the primary type of trailer used by 67.1% of haulers to transport cows and bulls to market, followed by gooseneck trailers at 30.3%. On average, haulers using pot belly trailers (n=95) used five compartments for separating cattle. Separating cattle using a center gate and compartments reduces carcass bruising and animal welfare concerns. It might seem that separating cows and heifers from bulls during transport might reduce the incidence of bruising; however, 64.4% of loads hauling both cows and bulls did not separate them according to sex (Harris et al., 2017).

5 Muscling and body condition of cull cows arriving at slaughter plants

Anecdotal information and findings of the 2016 National Beef Quality Audit (NBQA) indicate that a significant number of cull cows from the dairy industry are arriving at slaughter plants in less than satisfactory condition. As an example, the recent NBQA scored cattle for muscling as follows: score 1 = extremely light muscled to 5 = extremely heavy muscled. Statistics indicate that 93.8% of dairy cows were extremely light muscled or light muscled (score of 1 or 2); only 4.2% scored 3, 1.9% scored 4 and none scored 5. By comparison, only 34.7% of beef cows had extremely light or light muscling scores. Cattle with the greatest frequency of a muscle score of 3 were beef cows, beef bulls and dairy bulls (Harris et al., 2017).

Similar observations were made for body condition. Body condition scoring (BCS) is a useful management tool for determining nutritional needs and body reserves in cows. A 9-point scoring system is used for beef cows and beef bulls, whereas dairy cows and dairy bulls are scored using a 5-point scoring system with half scores. For perspective, a beef cow with a score of 5 or 6 is considered to be in ideal condition. A BCS of 1 is considered to be extremely thin and score of 9 would be considered a very obese cow. Using the 5-point scale with half scores for dairy cows and bulls provides a 9-point system similar to that for beef cattle. A cow with a BCS of 1 is considered a very thin cow and one with a BCS of 5 is considered very obese. Regardless of the stage of lactation, a cow with a BCS of 2 would be on the edge of being too thin and one with a BCS of 4 would be too fat. Lactating dairy cows scoring between 2.5 and 3.5 are considered to be within the ideal range.

The mean BCS recorded in the 2016 NBQA for beef cows (n = 1 911) and beef bulls (n = 406) was 4.7, whereas the mean BCS for dairy cows (n = 2 878) and dairy bulls (n = 121) was 2.6 and 3.3, respectively. Compared with data collected from 2007 (Nicholson, 2008) body condition of cows (both beef and dairy) improved from 2007 to 2016. Nonetheless, 9.3% of dairy cows and 0.8% of dairy bulls had BCS characterized as 'too low' (i.e. from 1.0 to 1.5). For the sake of comparison, 7.6% of beef cows and 5.7% of beef bulls had BCS considered too low (Harris et al., 2017).

The message to beef producers and dairy farmers is that when or if economically feasible, producers should strive to improve body condition and muscling in cattle determined to be light prior to marketing and/or slaughter. In some cases, marketing cattle while still in good condition, or increasing feed for light-muscled cattle and those with lower BCS has been shown to increase muscle and fat in thin under-conditioned animals (Matulis et al., 1987; Schnell et al., 1997). Improvements in body condition and muscling are also key determinants in market readiness, fitness for travel and welfare of cattle.

5.1 Physical defects observed in beef and dairy culls at slaughter

Based upon observations from the 2016 NBQA, the majority of cattle observed at slaughter facilities had no obvious injuries or physical defects. This suggests that most cattle were culled for reasons such as reproductive failure, undesirable behavioral characteristics or other non-physically obvious reasons. Physical defects that were observed in beef cows (*n*= 1 912), beef bulls (*n* = 402), dairy cows (*n* = 2 855) and dairy bulls (*n* = 120) included the following: bottle teats, broken penis, failed suspensory ligament, foot abnormality, full bag, lumpy jaw, mastitis, multiple udder problems, retained placenta, swollen joints and warts. At least one of these visible defects was present in 44.1%, 32.1%, 27.9% and 24.1% of dairy cows, beef bulls, beef cows and dairy bulls, respectively. Foot abnormalities (12.7%), swollen joints (4.7%) and the presence of an abscess (2.7%) were observed with the greatest frequency in beef bulls. The percentage of foot abnormalities in beef cows was 3.4% followed by 2.5% in dairy cows (Harris et al., 2017).

Chilean researchers recently reported on the pre- and post-transport effects on the welfare of 237 cull cows shipped directly from farms with an average journey length of 5 h 22 min. The shortest journey was 45 min and the longest was a maritime-terrestrial transport from a farm in the Magallanes Region that lasted 46 h 55 min. Cattle on the longest journey received neither water nor food at any time during the trip (Sánchez-Hidalgo et al., 2019). All cattle had at least one visible health problem. Fifty-two percent (124/237) of cattle had low body condition, 50% (119/237) mammary gland disorders and 24% (*n* = 57) were lame (Sánchez-Hidalgo et al., 2019).

6 Hide and carcass bruising evaluation

Branding reduces the value of cattle hides. It also causes acute as well as persistent pain (Tucker et al., 2014). Cattle that received a brand experienced pronounced pain in the days immediately following branding that lasted for 8 weeks following the date of procedure. Of cattle surveyed in the NBQA study, 22.7% had a least one visible brand. The percentage of beef cattle hides with brands was 35.7% compared to dairy cattle at 10.7%. Although hot-iron branding is the most permanent form of identification, it devalues the hide by as much as \$7.47 USD per animal (National Cattlemen's Beef Association, 2017) and presents a significant animal welfare concern. Better methods (less painful) of animal identification are needed.

6.1 Hide-off evaluation of carcass bruising

Carcass bruising may occur during any phase of the pre-transport, transport or post-transport process. Bruises require trimming that costs the cattle industry millions of dollars annually. In the 2016 NBQA, 35.9% of cattle (n = 4 262) had carcasses with no bruises, 67.3% bruises of minimal severity, 45.1% characterized as major, 4.9% critical and 1.4% as extreme. Observations in bulls (n = 389) revealed 57.5% with no bruises, 42.4% with minimal, and 21.9% with major bruises. When bruises are minimal in severity, they require minimal amounts (less than 0.45 kg) of surface trim. Comparing dairy with beef cattle, multiple bruises were observed in carcasses of 41.3% of dairy and 24.0% of beef cow carcasses. Similarly, 25.2% of dairy bull and 13.5% of beef bull carcasses had multiple bruises (Harris et al., 2017). At least part of the solution for reducing the risk of bruising is to improve body condition of cows and bulls.

In cow carcasses, the greatest percentage of bruises were located on the round and sirloin, whereas in bulls bruises occurred with greater frequency on the brisket, plate (behind the brisket and below the rib), and flank region when compared to cow carcasses. Bruises occurring within 24 h of harvest are usually a direct result of handling practices and facility design (Harris et al., 2017). Cows with poor body condition also had more severe bruising.

6.2 Predisposing factors and causes of bruising

Bruises are an indicator of poor animal welfare and an important cause of economic loss to packers (Grandin, n.d.; Huertas et al., 2010; Lee et al., 2017). The literature is replete with studies that demonstrate that bruising may occur at the farm, during transport and at slaughter (Harris et al., 2017; Grandin, n.d.; Jarvis et al., 1995; Sánchez-Hidalgo et al., 2019; Bethancourt-Garcia et al., 2019). It is important to identify the cause of bruises and at what point of the transport phase they may have occurred.

Pre-slaughter factors include the degree of muscling and BCS, gender and breed, horned versus polled cattle, facilities and handling at the farm of origin, and use of prods (Harris et al., 2017; Jarvis et al., 1995; Sánchez-Hidalgo et al., 2019; Bethancourt-Garcia et al., 2019). In the recent study by Sánchez-Hidalgo et al., poor body condition increased the likelihood of bruising by 20-fold.

Bruises occurring during transport may be a consequence of size and type of transport vehicle (i.e. larger versus smaller trucks) (Lee et al., 2017), animal density, the distance travelled and whether animals have been initially transported to markets or directly to slaughter. Inexperienced drivers or those who are prone to driving and taking corners too fast, making sudden stops or accelerating too rapidly may be to blame for an increase in bruising. When bruising is extensive, it may indicate that the animal had fallen and was trampled in transit (Tarrant et al., 1988; Grandin, https://www.grandin.com/references/cause.bruising.html). Holstein cattle experienced more traumatic events than beef breeds when hauled in potbelly trailers normally used for fat cattle and feeder calves. Researchers suggested that the larger frame scores/sizes were likely the predominant contributing factor (Lee et al., 2017).

If bruises are packing plant-related, carcasses will have bruises in the same anatomic location regardless of the origin of the cattle (Grandin, https://www.grandin.com/references/cause.bruising.html). Plants should inspect unloading ramps, alleyways, chutes and restrainers for evidence of broken parts or protruding objects that might cause bruising or injury. They must insure that employees understand proper animal handling techniques and places in the processing system where the potential for bruising may occur. Understanding the causes of bruising at the farm or feedlot, in transit or at the packing plant isthe first step in reducing the incidence of bruising and improving the welfare of cattle during transport and slaughter.

6.3 Dark cutting beef (DCF)

'Dark cutters' are a major meat-quality concern, but they are also indicative of welfare issues. It occurs in cattle that undergo a significant degree of pre-slaughter stress such as that associated with transport or rough handling at some point during the pre-slaughter process. Dark cutting beef (DCB) is described as dark red in color, lacking desirable tenderness, having an objectionable flavor and a short shelf-life. The dark color of the meat is caused by a depletion of glycogen stores in muscle tissues that raises the pH thereby facilitating bacteria growth on the meat's surface. In the absence of glycogen, the bacteria use glucose and amino acids as substrates for their metabolism and in the process generate putrefactive amines that give DCB its characteristic spoiled odor and objectionable taste.

An extensive review by Ponnampalam et al. (2017) indicates that on-farm contributing factors may include farm and feedlot feeding and management practices, animal genetics, breed and many others. Off-farm factors include marketing and transport conditions, lairage in route to and at the packing plant, and stunning and slaughter practices at the packing plant (Ponnampalam et al., 2017). It is beyond the scope of this chapter to discuss this issue in detail; instead, readers are directed to the review by Ponnampalam et al. for additional information on the subject of DCB.

7 Optimising welfare during the pre-transport stage

Transport is one the most stressful events cattle ever experience. It is compounded by the presence of disease or injury, recent weaning, co-mingling with unfamiliar animals, exposure to multiple handlers, inclement weather conditions, long-distance travel without feed or water and arrival to a completely new environment. Pre-planning at the pre-transport and transport phases along with proper handling and management of cattle at slaughter can minimize the negative impacts on end-product quality and animal welfare.

7.1 Pre-transport

The pre-transport phase includes decision-making on animals to be culled or removed from the farm or ranch. Once this is determined, the next assessment is the animal's fitness for transport be it to a livestock market or to slaughter. Animals that are strong, healthy and free of conditions affecting their mobility may be considered for sale at a livestock market or auction. Cattle with physical limitations (mild or moderate lameness, for example) yet still fit the requirements for human consumption should be sent to the nearest slaughter facility. Other procedures that may appreciably affect the welfare of cattle during the pre-transport phase are penning and handling at loading.

7.2 Determining fitness for travel

'It is impossible to assure good animal welfare during transport if the animal is unfit' (Cockram, 2019). Animals sent for slaughter with preexisting conditions are more likely to die in transit, become non-ambulatory, or require euthanasia on arrival to the plant (Cockram, 2019). In Canada, compromised cattle may be transported if they are sent directly to a nearby slaughter facility. They must be carefully loaded and segregated to avoid possible injury from interaction with other animals. They must be loaded last and unloaded first. Producers must weigh the potential for a financial return with the risks of animal suffering. They must consider the possibility of financial loss from mortality or condemnation of the carcass and the likelihood that an inspector or veterinarian may decide that the animal should not have been sent (Cockram, 2019).

The importance of carefully examining compromised animals prior to loading and movement to slaughter cannot be overemphasized. The Canadian Food Inspection Agency conducted a survey in 2001 to determine the number of non-ambulatory cattle arriving at federally inspected slaughter plants and auction markets in Canada. Of the 7382 non-ambulatory cattle arriving at these sites, 89.8% were dairy cattle with only a little over 10% identified as beef cattle (Doonan et al., 2003). It was further determined that less than 1% became non-ambulatory due to injury in transit or upon arrival at their destination. Following inspection, 37% of the non-ambulatory dairy cattle were condemned (Doonan et al., 2003).

Definitions for deciding fitness or unfitness for travel of compromised cattle are vague, subjective, and therefore inconsistent. A questionnaire survey of Danish livestock haulers found that 35% were frequently in doubt about specific cows when it came to their assessments regarding fitness for travel. When asked about transport legislation regarding fitness for travel, only 52% of drivers were able to provide correct answers to the questions (Herskin et al., 2017).

An animal's 'fitness for transport' generally refers to their ability to withstand the stresses of transportation such that it will not cause further compromise of their welfare. This would include animals that are non-ambulatory, suffering severe lameness, in poor body condition and/or otherwise in poor health that would suggest they are unlikely to be able to withstand long-distance travel to a market or slaughter plant.

Lactating dairy cattle are likely to become very uncomfortable under circumstances where they cannot be milked at least 2 times per day. They should not be transported long distances or sold through markets where it may be several days before they reach a slaughter plant. Ambulatory animals with elevated body temperatures, drug residue associated with recent treatment, obvious fractures, severe injury or open wounds, blindness, cancer (lymphoma), unreduced prolapses (uterine, rectal or other) or animals in labor with the potential to calve during transport do not fit the criteria of 'ready for transport' and should not be transported (AABP Guidelines for Transportation and Fitness-to-Travel Recommendations for Cattle https://www.aabp.org/Resources/AABP_Guidelines/transportationguidelines-2019.pdf).

The Ontario Farm Animal Council's 'Caring for Compromised Cattle' document (http://gpvec.unl.edu/Elective_files/feedlot/Caring_Compromised_Cattle_ISU.pdf) suggests that decisions regarding an animal's suitability for transport be determined by producers asking themselves three basic questions prior to loading animals for transport: (1) Can the animal walk? (2) Will the animal be able to walk off the truck at the final destination? (3) If the animal were to be moved to slaughter, would they themselves feel safe consuming meat from the animal? A 'no' response to any of these questions means that operators would need to consider further treatment or on-farm euthanasia. Treatment is a reasonable option when it is likely to improve the animal's condition. If treatment is unlikely to result in recovery and human consumption would pose risks to human health, euthanasia is the better choice. Euthanasia procedures should conform to the American Veterinary Medical Association's 2020 guidelines for euthanasia (AVMA Euthanasia Guidelines 2020, https://www.avma.org/sites/default/files/2020-01/2020_Euthanasia_Final_1-15-20.pdf). The overriding consideration in these decisions is always the welfare of the animal.

7.3 Pre-planning to reduce cattle hauling risks

The process of assembling, loading, hauling and unloading cattle provides plenty of opportunity for additional distress and injury. Co-mingling with other animals, handling by people unfamiliar to them and placement in new or strange environments all compound the challenges animals face during transport. Cattle with painful conditions such as lameness whether it be a foot problem, or proximal limb injury are likely to endure even greater pain when transported. Postural responses to side-to-side, back to front or front to back movement causes fatigue and pain on painful joints and muscles (Cockram, 2019).

Anecdotal estimates are that 95% of cattle are transported by truck in the United States. Beef and dairy cattle operations often haul cull cows and bulls to nearby markets or slaughter plants in pick-up pulled gooseneck trailers. On occasion, large dairies and cattle operations will remove a sufficient number of animals from their herds to fill a semi-tractor trailer. In the United States, most beef calves travel from cow/calf operations to livestock markets, then on to feeding areas several states away. In some cases, large cow/calf operations are able to sell calves direct to feeding operations, thus bypassing markets. Dairy calves and heifers are sometimes raised at off-site locations several states away from the home farm.

It is very inefficient and costly for livestock haulers to make short or certainly long hauls without being fully loaded. This may require that they plan additional stops to ranches, feedlots or markets when and where cattle are likely to be assembled and ready to be loaded. Haulers need to know the approximate size and numbers of animals to be loaded to assure they will have adequate space and be able to distribute cattle on the trailer properly. If a hauler is moving fat cattle to a slaughter facility, they may be required to load cattle at a specific time of day or night while simultaneously arranging their travel schedule to arrive at a packing plant at a specific time of day in order to fit the queue or slaughter sequence at the abattoir. As slaughterhouses become scarcer, and the demand for cost-effective hauling services increases, the need for trip planning to maximize efficiency becomes greater and more complicated. Poor logistical management that leads to increased transport time affects morbidity, mortality, weight loss, carcass bruising and quality. Minimizing transport time and travel distance are primary objectives for preserving animal welfare.

7.4 Pre-planning for transport during inclement weather

It is not always possible to predict the potential for encountering weather extremes particularly when traveling significant distances. However, pre-planning based upon the best available weather information is a key step in the pre-transport process. If there is high likelihood of poor weather (heavy snow, for example), haulers may need to consider rerouting or delaying the trip until conditions improve. Drivers should anticipate that it would take longer to get to the final destination and determine what that might mean for the animals in transit.

Transporting animals during periods of extreme cold weather is not recommended. If cattle must be transported in cold and windy conditions, drivers should avoid stopping in order to arrive at their destination as quickly as possible. At temperatures of −0°F (−18°C), hauling cattle at highway speeds risks dangerous wind chill. The likelihood of animal death increased sharply when the midpoint ambient temperature fell below −15°C (5°F) (Gonzalez et al., 2012). The problem is compounded when cattle are wet. Steps to mitigate the cold temperature effects on animals would include the following measures. Cover bottom ventilation slats in the vehicle to protect cattle from cold air drafts and road spray, but allow for adequate ventilation at all times. Close nose vents in the trailer, supply ample bedding such as straw and avoid overcrowding to allow space for animals to reposition themselves to avoid the direct effects of wind and cold that might cause frostbite. Wet bedding should be removed and replaced with clean and dry bedding materials after each trip (Master Cattle Transporter Guide, https://www.bqa.org/media/bqa/docs/master_cattle_transporter_guide-digital.pdf).

Extreme heat and humidity can be deadly as well. The likelihood of animals becoming non-ambulatory increased when temperatures rose above 30°C (86°F) (Gonzalez et al., 2012). Temperature-humidity indexes equal to or greater than 100°F (38°C) pose extremely serious if not deadly risks to cattle (Knowles, 1999). If possible, avoid transporting animals on hot and humid days. However, if necessary, transport of cattle should be avoided during daylight hours between 11 AM and 4 PM. Drivers should avoid stopping, but when required they should look for shaded areas and keep stops as short as possible. As with cold weather conditions, reduce cattle density by loading fewer animals on the trailer. Extreme care in handling is especially critical under hot and humid weather conditions (Master Cattle Transporter Guide, https://www.bqa.org/media/bqa/docs/master_cattle_transporter_guide-digital.pdf).

7.5 Assembling, handling and loading

The separation and re-corralling of animals for the purpose of confinement in a pen or lot will naturally cause some degree of angst and discomfort. It is compounded when animals are confined or regrouped with new animals from other sources. In lactating cows, regrouping reduces milk production and time spent feeding and resting. Even minor changes cause variable degrees of aggression normally exhibited as dominance behavior. For example, fighting, in the form of head-to-head pushing will often continue for a period of 2–3 days while a new hierarchy is established (Phillips, 2002). While difficult to prevent these types of interactions, they can result in injury and carcass bruising. Normal patterns of feed and water consumption are also disrupted which is problematic for animals that may be transported long distances where access may be denied for 24 h or more.

Loading areas should have smooth solid sides free of protrusions that might cause injury. Floors should have secure non-slip footing which may include the use of rubber mats, sand and textured concrete surfaces. Handle cattle quietly during the loading process in order to avoid bruising or injury from slipping and falling. Chutes, alleyways, ramps and races should have secure footing, and solid sides free of protruding objects that could cause bruising. Loading ramps should be at an angle of less than 25° with non-slip flooring and no gaps between the ramp and the floor of the trailer.

Cull dairy cows that are lactating should be milked just prior to loading and transport. Be sure cattle have access to feed and water prior to transport to a market or distant packing plant and assure appropriate density on the trailer. Consider weather conditions and adjust animal density as necessary. Do not load cattle with conditions that are unlikely to pass ante-mortem inspection at slaughter including cancer eye and blindness, fever above 103°F (39°C), unreduced prolapses, severe emaciation and weakness, or cows in labor or where calving is imminent. Cows with lameness should be carefully scrutinized. Do not load cows with a locomotion score of 4 or 5 on a 5-point scale; rather consider the options of additional treatment or euthanasia.

8 Welfare of cattle during transport

8.1 Transport time and distance

The effect of the time a journey requires appears to be of greater significance than distance covered (Knowles, 1999; Gonzalez et al., 2012). Canadian researchers assessed the effect of long haul transport (> 400 km, i.e. 250 miles) on the welfare of cattle (Gonzalez et al., 2012). The incidence of dead, non-ambulatory and lame cattle, characterized as 'total compromised' animals, was determined using survey responses (n= 6,152) from drivers and hauling companies (16), involving 327 tractor-trailer units and 290 866 animals. The total number of compromised animals was 130 (0.045%–130/290 866); 32 (0.011%) died, 65 (0.022%) became non-ambulatory and 34 (0.012%) became lame in transit.

Other findings from this study were that cull cattle were at greatest risk of becoming lame at the time of loading and unloading, becoming non-ambulatory or dying during the journey compared with feeder calves and fat cattle. Observations also confirmed that the likelihood of death, becoming lame or non-ambulatory increased the longer animals were in transit. Death rate was higher for animals that lost more than 8% of body weight. The high loss of body weight was attributed to ambient air temperatures above 30°C (86°F) and a long period of time (> 30 h) on the truck (Gonzalez et al., 2012).

8.2 Stocking density

Despite industry proponents who argue for high stocking densities, research indicates that high stocking densities (Tarrant et al., 1988) reduce the welfare of animals during transport. When stocking densities are low and movement is less restricted, cattle are able to change positions frequently during the trip. The preferred orientation of cattle in low-density conditions is parallel to the direction of travel and loss of balance occurs primarily on gear changes, braking and cornering. In high stocking density conditions, movement is more restricted, thereby interfering with the animal's ability to adopt a preferred orientation to the direction of travel. Advocates of high stocking densities suggest that the opportunity to lean on an adjacent animal counteracts minor losses of balance, creating less potential for falling during transport.

The primary hazard in transport is for cattle that loose footing and go down during transport. Involuntary losses in balance that led to falling occurred more frequently in high stocking density conditions (Tarrant et al., 1988). Moreover, animals that went down in the trailer were subject to trampling and destabilization of the entire group that on occasion led to what authors' described as a 'domino effect'. Fear of falling is believed to be the primary reason cattle stand during transport and are likely to lie down only on very long journeys (Knowles, 1999). Overall, animal welfare and carcass quality are more adversely affected at higher stocking densities compared with medium and low stocking densities (Tarrant et al., 1988).

8.3 Behavior of cattle during transport

Cattle are naturally anxious and restless at the beginning of their journeys, which leads to frequent bouts of urination and defecation. Manure slurry and moisture reduces traction on trailer flooring surfaces and can increase the potential for falling. Social interactions characterized as head butting, mounting and pushing behavior are more commonly observed soon after loading, during the early part of the journey and when stocking densities are higher (Tarrant et al., 1988; Knowles, 1999). Exploratory behavior described as sniffing the environment, sexual and aggressive behaviors were reduced in high stocking density conditions.

8.4 Drivers and driving experience

Anyone who might think the job of a trucker who hauls a load of lumber is equivalent to hauling a load of livestock, should think again. Transporting live animals safely and on time to their destination can be extremely challenging. Livestock haulers must understand how to handle cattle and load a truck and trailer to distribute the weight appropriately. They must abide by schedules for cattle pickups at feedlots, farms as well as markets, and time these so that they can meet delivery schedules whether it be to another farm, market or to slaughter. Feedlot operators pay for losses that occur because of 'shrink'; therefore, producers will select haulers that can get cattle to packing plants with a minimum of weight loss. The health and welfare of cattle during transport is largely determined by the ability and experience of the driver.

Mortality rates for cattle during road transport are low compared to other species (Phillips, 2002; Gonzalez et al., 2012). More important are bruises that occur as a result of injuries or bruising associated with vehicle movement. Driving and driver experience are the most important factors with respect to injuries and bruising during transport (Knowles, 1999). Rapid starts, quick acceleration, braking, gear changes and excessive speeds while cornering are all causes for instability that increase the potential for bruising and in worst-case scenarios falling. Selecting routes that minimize the number of hills and winding roads can reduce these issues significantly.

In a long haul study by Gonzalez et al., the driver's experience had a significant impact on the welfare of cattle. Drivers with greater than 10 years of experience had a lower incidence of non-ambulatory and compromised cattle on arrival at their destinations compared to drivers of less than 10 years of experience. It is likely that experienced drivers not only have better driving skills, but also do a better job of distributing animals in the trailer. Researchers suggested that it may be advantageous to use experienced drivers when encountering weather extremes and journeys lasting more than 30 h (Gonzalez et al., 2012).

8.5 Livestock rollover accidents

There are few published reports regarding highway accidents involving livestock haulers. One that is frequently sighted is that by Woods and Grandin, 2008 (Woods and Grandin, 2008) of 415 commercial livestock truck accidents occurring in the United States and Canada between 1994 and June 2007. Fifty-nine percent of the accidents from this report occurred during the early morning hours from midnight to 9 AM. The majority, 80% of the accidents, involved a single vehicle. Driver error was cited as the cause for 85% of the accidents. In most cases (84%) the vehicle rolled over to the right side. Since vehicles travel on the right side of the road in North America, it was deduced that most cases result from the driver's fatigue and falling asleep while driving. Accidents were most common in the fall of the year and 69% of the trucks were hauling calves and yearlings to grower or feeder operations in excess of 500 km (310 miles). Most trips are made by a single driver with travel times of 20 h or more (Woods and Grandin, 2008).

8.6 The 28-hour law

The goal should always be to find markets and slaughter plants as close to the farm as possible. Journey time should be short enough to avoid the requirement to unload the animals for resting, which in the United States is 28 h. The 28-h law provides that animals cannot be transported by 'rail carrier, express carrier or common carrier' (except by air or water) for more than 28 consecutive hours without being unloaded for 5 h for rest, water and food (United States Code Annotated. Title 49. Transportation. Subtitle X. Miscellaneous. Chapter 805. Miscellaneous. § 80502. Transportation of animals. https://www.govinfo.gov/content/pkg/USCODE-2011-title49/pdf/USCODE-2011-title49-subtitleX-chap805-sec80502.pdf). If transport is anticipated to be equal to or more than 28 h, haulers must plan their routes so that rest periods may be accommodated. This rule, as good as it seems to be in principle, has been difficult if not impossible to enforce (Edwards-Callaway et al., 2019). Although it would seem logical to expect that a stop along the way in long-distance travel would be of value, there is also very little if any evidence to support benefits or reasonable guidelines.

8.7 Hours-of-service driving law

This law (developed during the Obama Administration in 2015) requires that truck drivers not be officially on duty for more than 14 h and while on duty, not have driven more than 11 of those hours. When drivers reach those maximum allotments of time, they are required to stop and rest for a minimum of 10 h. To assure compliance with the law, in December 2017 the Federal Motor Carrier Safety Administration (FMCSA) began requiring most US truckers to carry electronic logging devices (ELD) to verify that truckers were stopping to rest as required. The FMCSA, yielding to pressures from the agricultural commodities granted drivers an exemption from the requirement to use ELDs under certain travel conditions. As the law is currently implemented, cattle haulers are subject to the hours of service regulations when operating beyond 150 air miles from the source of the commodity; that is 11 h of driving, 14 h of on duty not driving and 10 consecutive hours off before proceeding to drive for 11 h. While operating within said 150 air miles, the hours of service rules do not apply. Drivers transporting livestock or insects are not subject to the ELD rule (https://cms8.fmcsa.dot.gov/regulations/hours-of-service).

8.8 The 'Transporting livestock across America safely Act'

On January 10, 2019, Representative Ted Yoho (also a veterinarian) from Florida, introduced bill H.R. 487. 'Transporting Livestock Across America Safely Act' in the House of Representatives to develop a ruling that would be more acceptable to livestock haulers across the United States. Soon after its introduction to the House of Representatives, this bill was referred to the Subcommittee on Highways and Transit in February 2019 where it awaits further action. Among other provisions, this bill increases the drive time to a maximum of not less than 15 h and not more than 18 h within a 24-hour period. It also excludes time spent loading and unloading to more accurately include only travel time, which should make the rule more reasonable for livestock haulers (https://www.congress.gov/bill/115th-congress/senate-bill/2938/text?format=txt). No doubt, if this bill were likely to become law, it will undergo significant modification; nonetheless, efforts to move animals to their destinations quickly and safely are important for human health and animal welfare.

The 28-h law, 'Hours of Service Law' and 'Transporting Livestock Across America Safely Act' are intended to reduce the number of accidents involving over-the-road haulerswhile not causing unreasonable restrictions for livestock haulers. Nationwide, accident statistics indicate that truck-related fatalities have been on the rise in recent years. For example, in 2017, there were 841 occupants of large trucks killed in crashes, which was up from 725 in 2016, and 665 in 2015, according to a National Highway Traffic Safety Administration report Traffic Safety Facts 2017 Data – Large Trucks https://crashstats.nhtsa.dot.gov/Api/Public/ViewPublication/812663. The most common causes of truck accidents are fatigue, impaired and distracted drivers. Meaningful laws that can promote highway safety are good for humans and animals alike.

9 Welfare of cattle at slaughter

The welfare of cattle at slaughter is determined by the evaluation of animal-based measures assessed on arrival, while in lairage, during movement to the stunner and at the point of stunning, exsanguination and death. Specific outcome measures include an enumeration of the following: animals dead or non-ambulatory on arrival; the number of animals observed to slip or fall during unloading, in lairage, or during movement to the stunner; evidence of vocalization, amount of prod use, time from stunning to exsanguination ('stun to stick' time) and effectiveness of stunning procedures.

9.1 Unloading

The first step on arrival at a packing plant is unloading. Under ideal conditions, this should occur within 15 min and hopefully no more 60 min after arrival.

Trucks are normally scheduled so that delays are minimized which is critically important in hot weather to avoid heat stress. Cattle should exit the truck or trailer calmly with little or no encouragement. Industry guidelines recommend that less than 5% of animals be unloaded with the use of electric prods and less than 1% of animals fall during the unloading process (Grandin, 2005a, https://www.grandin.com/RecAnimalHandlingGuidelines.html). Unloading areas should have secure flooring surfaces and provide a sufficiently wide and clear path to the holding pen or lairage area. Concrete surfaces may be grooved in a series of parallel grooves or in a diamond pattern to increase traction on concrete floors. Cattle should be carefully observed at unloading for evidence of reduced mobility or other condition (2016 Guidelines for Humane Slaughter, https://www.avma.org/sites/default/files/resources/Humane-Slaughter-Guidelines.pdf).

9.2 Mortalities and non-ambulatory cattle on arrival at slaughter

The number and percentage of animals dead or non-ambulatory on arrival at slaughter plants should be monitored. Both conditions are most likely the result of the shipment of animals that were not fit for travel. In the United States, non-ambulatory disabled cattle are defined as any animal that cannot rise from a recumbent position or cannot walk. Such cattle cannot enter a packing plant. This includes animals with broken legs, severed tendons or ligaments, paralysis from nerve injury, and fractures of the vertebral column or metabolic conditions. This also applies to cattle demonstrating neurological symptoms even if they are ambulatory.

At one time, the Food Safety and Inspection Service (FSIS) would permit the slaughter of cattle under 'emergency conditions'. For example, an animal that had passed ante-mortem inspection, but later sustained an acute injury in route to the knock box (cattle stunner) could still be slaughtered. The agency reversed its position on emergency slaughter in early 2008 following the inhumane handling of non-ambulatory animals at a California slaughter plant. That incident resulted in the recall of 143 million pounds of fresh and frozen beef products dating back to February 2006 (the largest meat recall in US history). Regulations now prohibit the slaughter of any non-ambulatory animal arriving at a slaughter plant (Becker, 2009 https://nationalaglawcenter.org/wp-content/uploads/assets/crs/RS22819.pdf .

9.3 Severe lameness and emaciation

Regulations on slaughter of cattle also extend to animals with severe lameness and emaciation. Nicholson et al. (2013), observed that 2.7% of cull dairy cows were severely lame (i.e. displaying an arched back at all times and refusing to bear weight on one leg), 4% of cattle were extremely emaciated and 4% had foot abnormalities (Nicholson et al, 2013). If an animal is too weak or lame to freely walk off the truck, it will likely have difficulty walking to the stunner and is therefore unlikely to pass ante-mortem inspection. A survey of 10 cattle auction markets in the western United States, found that 15.1% beef cows, and 15.4% of beef bulls were lame. Observations were worse on dairy cows, with lameness found in 44.7% of dairy cows and 26.1% of dairy bulls. This survey also found severe emaciation (i.e. dairy BCS = 1 and beef BCS = 1) in 13.3% of cull dairy and 3.9% of cull beef cows (Ahola et al., 2011).

9.4 Lairage

Cattle should be maintained with the animals in which they arrived. This is necessary to reduce the potential for fighting and dominance behavior that increases stress in cattle and the amount of carcass bruising and dark cutting beef. Bulls are prone to fighting and mounting and may need to be separated to avoid excessive bruising and dark cutters. Lairage pens should be of sufficient size to accommodate a full truckload of cattle. Pens should have water troughs and sufficient space so that animals are able to lie down, rest and allow observation by veterinary inspectors during ante-mortem inspection (AVMA Humane Slaughter Guidelines, 2016).

9.5 Movement to the stunner

As cattle move from lairage pens to the lead up alley heading to the stunning area, there should be no slipping, falling or vocalization. Alleyways should be free of extraneous debris that might cause distraction and hesitation in the forward movement of cattle. Animals prefer to go from dark areas to areas that are well lit; therefore, lighting that eliminates shadows improves cattle flow and reduces balking. Animal handlers must understand the principles of flight zone and point of balance for low stress handling of cattle. Handlers should also limit the number of animals in crowd pen areas to no more than 3–5.

Plants with good handling practices should be able to have less than 1% of animals falling and less than 3% vocalizing from adverse handling. Use of electric prods should be minimized by restricting use and availability of the devices to trained personnel and only in areas where animals are likely to balk (i.e. prior to entering the stun box) (AVMA Guidelines for Humane Slaughter 2016).

9.6 Restraint at the stun box

In packing plants where a stun box is used, pressure used to restrain the animal should be sufficient for restraint but not to the extent, that it causes pain or distress. Cattle are prone to balk upon entry to the stunning box, which may require the need for a prod or push gate. Improved designs move the front of the stun box assembly, including the head holder, backward toward the animal. This eliminates the need to prod or push the animal forward. Large plants use a center V track restrainer system, which maintains animals in an upright comfortable position while moved to the stunner on a conveyor. These systems also facilitate stunning by positioning the stunner operator in a safer and more effective position for proper anatomic placement of the captive bolt (AVMA Guidelines for Humane Slaughter, 2016).

9.7 Captive bolt stunning

The physical methods used for the humane stunning of cattle are penetrating and non-penetrating captive bolt and firearms. The objective of these devices is to cause an immediate loss of consciousness by causing trauma to the cerebral hemispheres and brainstem. Standing animals collapse immediately with a properly placed shot, followed by muscular rigidity and eventually relaxation that leads to a period of uncoordinated limb movement. There is an immediate cessation of rhythmic breathing and loss of corneal and palpebral reflexes.

There are two types of captive bolt, penetrating and non-penetrating. Penetrating captive bolt discharges a bolt through the skull and into the brain. Non-penetrating captive bolt is equipped with a convex mushroom-shaped head that causes unconsciousness by concussion, fracture of the skull and focal damage to the brain around the site of impact. Captive bolts are powered by either compressed air or gunpowder cartridge charges ranging from 9 mm, 22 caliber or 25 caliber. Dysfunction of powder-charged captive bolt guns is primarily due to a lack of maintenance or damp cartridges from improper storage of powder charges (AVMA Guidelines for Humane Slaughter, 2016).

Properly applied by a well-trained operator, captive bolt can easily render 95% or more of animals unconscious with a single shot (Grandin, 2002). Shot effectiveness below 95% is indicative of a problem with either the person doing the stunning or the stunning equipment. Grandin reports that the best plants have a 99% first shot efficacy (Grandin, 2005b). Head holders can be used to improve shot accuracy, but they must be properly designed or they may cause additional stress on restrained animals. The head remains stable without restraint on center-track conveyor systems, which helps to increase the accuracy of shot placement (2016 AVMA Humane Slaughter).

9.8 Stun to stick interval

Captive bolt stunning is intended to cause an immediate and sustained loss of consciousness persisting until exsanguination reduces blood pressure and volume sufficient to cause death. Prolongation of the time from stunning to exsanguination (better known as the 'stun-to-stick interval') risks the possibility of an animal regaining consciousness. The recommended maximum stun-to-stick times for cattle are 60 s for penetrating and 30 s for non-penetrating captive bolt (Captive-Bolt Stunning of Livestock (2013) Humane Slaughter Association, United Kingdom: https://www.hsa.org.uk/downloads/publications/captiveboltstunningdownload.pdf).

Exsanguination of sheep and goats is typically performed by making a ventral incision in the neck severing the carotid arteries and jugular veins. In cattle, exsanguination is performed by inserting a sharp knife into the jugular furrow at the base of the neck. Once the knife is through the skin it is directed toward the thoracic inlet where severing of the large arteries and veins at the base of the neck assures rapid loss of blood and death (Captive-Bolt Stunning of Livestock (2013) Humane Slaughter Association, United Kingdom: https://www.hsa.org.uk/downloads/publications/captiveboltstunningdownload.pdf).

9.9 Firearms

Gunshot is the most practical for slaughter of animals with heavy skulls such as large bulls, boars and bison. Safety is the major concern when firearms are used for slaughter or euthanasia purposes. Selection of a firearm and bullet that is sufficient to traverse the skull, but not perforate (i.e., pass through) the skull is desired. A properly placed bullet will cause massive brain damage and death with little likelihood of a return to consciousness.

Handguns in calibers ranging from .32 to .45 caliber, rifles from .22 magnum or higher for thick skulled bulls, boars and bison and shotguns gauges .410, 28, 20, 16 or 12 are acceptable. When using a handgun, rifle or shotgun at close range, hold the muzzle of the firearm within 3 feet of the intended target. Solid point bullets or hollow points with controlled expansion properties are best for handguns and rifles. Number 6 or larger birdshot is acceptable for shotguns; buckshot or slugs are acceptable but probably excessive since most shots are within close range (Shearer, 2018). Full metal-jacket bullets are not recommended for use in slaughter plants for safety reasons (AVMA Guidelines for Humane Slaughter, 2016).

An important difference with the use of a firearm as compared with captive bolt is that the muzzle of a firearm should never be held flush with the skull. Enormous pressures are developed within the barrel when it is discharged that could cause explosion of the barrel putting the shooter and by-standers at great risk. Captive bolt on the other hand, requires immediate contact with the skull otherwise, one is unlikely to realize the full benefit of bolt extension into the skull.

9.10 Anatomic landmarks for the use of the penetrating captive bolt and gunshot

Selection of the ideal anatomic site for use of a captive bolt or firearm is a key prerequisite to stunning success. In cattle, the point of entry of the projectile should be at the intersection of two imaginary lines, each drawn from the outside corner of the eye to the center of the base of the opposite horn (2020 AVMA Euthanasia Guidelines; AVMA Guidelines for Humane Slaughter, 2016). Alternatively, in long-faced cattle or young-stock, a point on the midline of the forehead that is halfway between the top of the poll and an imaginary line connecting the outside corners of the eyes can be used (2016 AVMA Humane Slaughter Guidelines). Firearms should be positioned so that the muzzle is perpendicular to the skull to direct the projectile toward the brainstem (Fig. 3).

Proper anatomic site selection and use of the appropriate caliber handgun, rifle or gauge of shotgun and bullets/shells assures effective results. A well-trained shooter should be able to effectively assure unconsciousness and death

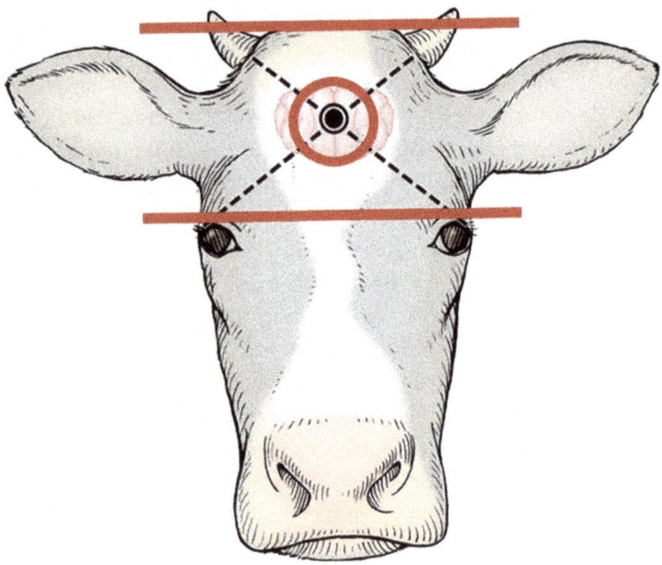

Figure 3 Guidelines for anatomic site placement of a captive bolt or bullet from a firearm. The point of entry of the projectile should be at the intersection of two imaginary lines (represented as the black dashed lines), each drawn from the outside corner of the eye to the center of the base of the opposite horn. This site can also be determined as a point on the midline of the face (defined by the red circle) that is halfway between the top of the poll and an imaginary line connecting the outside corners of the eyes can be used (adapted from the website of Shearer and Ramirez, Humane Euthanasia of Livestock, Iowa State University; AVMA Guidelines for the Humane Slaughter of Animals: 2016 Edition; AVMA Guidelines for the Euthanasia of Animals: 2020 Edition).

in 95% or more of the animals with a single shot (AVMA Guidelines for the Humane Slaughter of Animals: 2016 Edition https://www.avma.org/sites/default/files/resources/Humane-Slaughter-Guidelines.pdf accessed on January 2020.)

10 Summary

Transport is an extremely stressful event for cattle, but there are ways to reduce its impacts on their welfare. Dairy cattle typically leave herds at less than 5 years of age, whereas beef cattle are roughly 10 or more years of age. Most dairy and beef herds sell cattle through livestock markets. Compromised cattle should be transported directly to the nearest slaughter facilities. Dairy cattle arriving at slaughter facilities are frequently light muscled and in poor body condition. This increases the potential for bruising that causes greater economic loss from the need to trim away bruised tissues. The incidence of visually obvious physical defects in cattle is low; however, more than one-third of hides are damaged by branding. Carcass evaluation at slaughter plants demonstrates a high incidence of carcass bruising and dark cutting beef, in cases where animals have experienced a significant degree of pre-slaughter stress. These observations confirm the fact that transport and handling at slaughter are major welfare issues. Improving the welfare of animals and quality of beef from animals during transport and at slaughter begins at the pre-transport phase. Determining fitness for travel and pre-planning that takes into consideration the condition of animals to be transported, time and distance to the desired destination and weather conditions are key steps in optimizing welfare of animals that must be transported. The natural behavior of cattle is to stand while in transit. In long-distance travel, this increases fatigue and potential for injury, particularly for animals compromised by lameness, emaciation or other condition. Cattle should be carefully observed upon arrival at slaughter. If an animal is non-ambulatory it must be humanely euthanized. Animals that exit the trailer freely should be moved to lairage that offers immediate access to water and sufficient space for animals weary from travel an opportunity to lie down and rest. The plant's slaughter schedule should be managed such that cattle spend minimal time in lairage prior to movement to stunning and slaughter. Animals should be moved calmly and with a minimum of prodding in route to stunning and slaughter. Upon entry into the stun box, stunning should proceed without interruption. Personnel delegated the task of stunning should be able to effectively render animals unconscious with 95% or better accuracy with the first shot. If a second shot is required, it should be applied immediately. Facilities and movement of animals from the stun box to the bleeding rail should be designed to assure a maximum of 60 s from stunning to the point of exsanguination. Use of the appropriate anatomic site assures optimal success and animal welfare at slaughter.

11 Where to look for further information

There is an abundance of information available on the subject of livestock transport. Several key references for those desiring additional information are included in the list of references that accompany this article. However, consolidation of the livestock industry and the reduction in availability of livestock market and packing plants throughout North America will continue to create challenges for ensuring the welfare of animals that require transport. Interest in the ability to improve pre-transport preparation, to reduce stress during transport and to better manage conditions occurring in the post-transport phase will continue to be priorities for future research.

There are multiple organizations with a stake in assuring the welfare of cattle during transport including the National Cattlemen's Beef Association (NCBA), Livestock Market Association (LMA), National Milk Producers Federation (NMPF) and the North American Meat Institute (NAMI) to mention just a few. However, beyond producer, market and meat packers are consumers. Now, more than at any time in history, consumers want to know that the welfare of animals is assured on the farm, at markets at slaughter and points in between.

12 References

Ahola, J. K., Foster, H. A., VanOverbeke, D. L., Jensen, K. S., Wilson, R. L., Glaze, J. B., Fife, T. E., Gray, C. W., Nash, S. A., Panting, R. R. and Rimbey, N. R. 2011. Survey of quality defects in market beef and dairy cows and bulls sold through livestock auction markets in the Western United States: I. Incidence rates. *J. Anim. Sci.* 89(5):1474–1483.

American Association of Bovine Practitioners. 2019. AABP Guidelines for Transportation and Fitness-To-Travel Recommendations for Cattle. Available at: https://www.aabp.org/Resources/AABP_Guidelines/transportationguidelines-2019.pdf. Accessed January 2020.

American Veterinary Medical Association. 2016. AVMA Guidelines for the Humane Slaughter of Animals: 2016 Edition. Available at: https://www.avma.org/sites/default/files/resources/Humane-Slaughter-Guidelines.pdf. Accessed January 2020.

American Veterinary Medical Association. 2020. AVMA Guidelines for the Euthanasia of Animals: 2020 Edition. Available at: https://www.avma.org/sites/default/files/2020-01/2020_Euthanasia_Final_1-15-20.pdf. Accessed January 2020.

Becker, G. S. 2009. *Nonambulatory Livestock and the Humane Methods of Slaughter Act*. Congresional Research Service. 7-5700, www.crs.gov, RS22819. p. 1-7.

Bethancourt-Garcia, J. A., Vaz, R. Z., Vaz, F. N., Silva, W. B., Pascoal, L. L., Mendonça, F. S., Vara, C. Cd, Nuñez, A. J. C. and Restle, J. 2019. Pre-slaughter factors affecting the incidence of severe bruising in cattle carcasses. *Livest. Sci.* 222:41–48.

Cockram, M. S. 2019. Fitness of animals for transport to slaughter: a Review. *Can. Vet. J.* 60(4):423-429.

De Vries, A. 2013. Cow longevity economics: the cost benefit of keeping the cow in the herd. Proceedings of the From the Cow Longevity Conference 2013, Hamra Farm, Sweden. Available at: http://www.milkproduction.com/Library/Scientific-articles/Management/Cow-longevity-economics-The-cost-benefit-of-keeping-the-cow-in-the-herd/. Accessed December 2019.

Doonan, G., Appelt, M. and Corbin, A. 2003. Nonambulatory livestock transport: the need for consensus. *Can. Vet. J.* 44(8):667-672.

Edwards-Callaway, L. N., Walker, J. and Tucker, C. B. 2019. Culling decisions and dairy cattle welfare during transport to slaughter in the United States. *Front. Vet. Sci.* 5: :1-5. Article 343.

Geiger, C. 2016. What is the national cull rate? Hoard's Dairymen Intel. Available at: https://hoards.com/article-19040-what-is-the-national-cull-rate.html. Accessed December 2019.

Gonzalez, L. A., Schwartzkopf-Genswein, K. S., Bryan, M., Silasi, R. and Brown, F. 2012. Relationships between transport conditions and welfare outcomes during commercial long haul transport of cattle in North America. *J. Anim. Sci.* 90(10):3640-3651.

Grandin, T. 2002. Return-to-sensibility problems after penetrating captive bolt stunning of cattle in commercial beef slaughter plants. *J. Am. Vet. Med. Assoc.* 221(9):1258-1261.

Grandin, T. 2005a. Edition, with 2007 and 2010 Updates. Recommended Animal Handling Guidelines and Audit Guide for Cattle, Pigs, and Sheep. Available at: https://www.grandin.com/RecAnimalHandlingGuidelines.html. Accessed January 2020.

Grandin, T. 2005b. Maintenance of good animal welfare standards in beef slaughter plants by use of auditing programs. *J. Am. Vet. Med. Assoc.* 226(3):370-373.

Grandin, T. n.d. How to Track Down the Cause of Bruising. Available at: https://www.grandin.com/references/cause.bruising.html. Accessed December 26, 2019.

Grandin, T. 2013. *Recommended Animal Handling Guidelines and Audit Guide: A Systematic Approach to Animal Welfare*. AMI Foundation, Washington, DC. Available at: https://www.grandin.com/RecAnimalHandlingGuidelines.html. Accessed 20 December 2019.

Harris, M. K., Eastwood, L. C., Boykin, C. A., Arnold, A. N., Gehring, K. B., Hale, D. S., Kerth, C. R., Griffin, D. B., Savell, J. W., Belk, K. E., Woerner, D. R., Hasty, J. D., Delmore, R. J., Martin, J. N., Lawrence, T. E., McEvers, T. J., VanOverbeke, D. L., Mafi, G. G., Pfeiffer, M. M., Schmidt, T. B., Maddock, R. J., Johnson, D. D., Carr, C. C., Scheffler, J. M., Pringle, T. D. and Stelzleni, A. M. 2017. National beef quality audit–2016: transportation, mobility, live cattle, and carcass assessments of targeted producer-related characteristics that affect value of market cows and bulls, their carcasses, and associated by-products. *Translat. Anim. Sci.* 1(4):570-584.

Herskin, M. S., Hels, A., Anneberg, I. and Thomsen, P. T. 2017. Livestock drivers' knowledge about dairy cow fitness for transport – a Danish questionnaire survey. *Res. Vet. Sci.* 113:62-66.

Hours of Service Rule. 2020. U.S. Department of Transportation. Federal Motor Carrier Safety Administration. Available at: https://cms8.fmcsa.dot.gov/regulations/hours-of-service. Accessed January 2020.

HSA. 2013. Captive-Bolt Stunning of Livestock. Humane Slaughter Association, United Kingdom. Available at: https://www.hsa.org.uk/downloads/publications/captivebo ltstunningdownload.pdf. Accessed January 2020.

Huertas, S. M., Gil, A. D., Piaggio, J. M. and van Eerdengurg, F. J. C. M. 2010. Transportation of beef cattle to slaughterhouses and how this relates to animal welfare and carcase bruising in an extensive production system. *Anim. Welf.* 19:281–285.

Jarvis, A. M., Selkirk, L. and Cockram, M. S. 1995. The influence of source, sex class and pre-slaughter handling on the bruising of cattle at two slaughterhouses. *Livest. Prod. Sci.* 43(3):215–224.

Knowles, T. G. 1999. A review of the road transport of cattle. *Vet. Rec.* 144(8):197–201.

Lee, T. L., Reinhardt, C. D., Bartle, S. J., Vahl, C. I., Siemens, M. and Thomson, D. U. 2017. Assessment of risk factors contributing to carcass bruising in fed cattle at commercial slaughter facilities. *Transl. Anim. Sci.* 1(4):489–497.

Livestock Trailer Safety. 2012. Farm and Ranch eXtension in Safety and Health (FReSH) Community of Practice. Available at: http://www.extension.org/pages/64391/livestock-trailer-safety. Accessed December 2019.

Master Cattle Transporter Guide, National Beef Quality Assurance Guide for Cattle Transporters. Available at: https://www.bqa.org/Media/BQA/Docs/master_cattle_transporter_guide-digital.pdf. Accessed December 2019.

Matulis, R. J., McKeith, F. K., Faulkner, D. B., Berger, L. L. and George, P. 1987. Growth and carcass characteristics of cull cows after different times-on-feed. *J. Anim. Sci.* 65(3):669–674.

National Cattlemen's Beef Association. 2017. *National Beef Quality Audit-2017: Market Cows and Bulls, Executive Summary*. National Cattlemen's Beef Association, Centennial, CO.

National Highway Traffic Safety Administration report Traffic Safety Facts 2017. Data - Large Trucks. National Center for Statistics and Analysis. (2019, January). Large trucks: 2017 data. Washington, DC: National Highway Traffic Safety Administration. Available at: https://crashstats.nhtsa.dot.gov/Api/Public/ViewPublication/812663. Accessed January 2020.

Nicholson, J. D. W. 2008. National Market Cow and Bull Beef Quality Audit-2007: a survey of producer-related defects. M.S. Thesis, Texas A&M University, College Station.

Nicholson, J. D. W., Nicholson, K. L., Frenzel, L. L., Maddock, R. J., Delmore, R. J., Lawrence, T. E., Henning, W. R., Pringle, T. D., Johnson, D. D., Paschal, J. C., Gill, R. J., Cleere, J. J., Carpenter, B. B., Machen, R. V., Banta, J. P., Hale, D. S., Griffin, D. B. and Savell, J. W. 2013. Survey of transportation procedures, management practices, and health assessment related to quality, quantity, and value for market beef and dairy cows and bulls. *J. Anim. Sci.* 91(10):5026–5036.

Ontario Farm Animal Council. 2010. The Ontario Farm Animal Council's "Caring for Compromised Cattle" document. Available at: http://gpvec.unl.edu/Elective_files/feedlot/Caring_Compromised_Cattle_ISU.pdf. Accessed January 2020.

Phillips, C. 2002. *Cattle Behaviour and Welfare* (2nd edn.), *Chapter 5: The Welfare of Cattle during Transport, Marketing and Slaughter*. Wiley-Blackwell, Hoboken, NJ. p. 38–48.

Ponnampalam, E. N., Hopkins, D. L., Bruce, H., Li, D., Baldi, G. and Bekhit, A. E. 2017 Causes and contributing factors to "Dark Cutting" meat: current trends and future directions: a review. *Compr. Rev. Food Sci. Food Saf.* 16(3):400–430.

Sánchez-Hidalgo, M., Rosenfeld, C. and Gallo, C. 2019. Associations between pre-slaughter and post-slaughter indicators of animal welfare in cull cows. *Animals* 9(642):1–15.

Schnell, T. D., Belk, K. E., Tatum, J. D., Miller, R. K. and Smith, G. C. 1997. Performance, carcass, and palatability traits for cull cows fed high-energy concentrate diets for 0, 14, 28, 42, or 56 days. *J. Anim. Sci.* 75(5):1195–1202.

Shearer, J. K. 2018. Euthanasia of cattle: practical considerations and application: a review. *Animals* 8(57):1–17.

Stalsberg, K., Halfman, B., Stuttgen, S. and Skjolaas, C. A. 2017. *Livestock Hauling BQA Style, Wisconsin Beef Quality Assurance*. University of Wisconsin-Extension. Available at: https://fyi.extension.wisc.edu/wbic/files/2018/03/3T-handout-1done .pdf. Accessed December 2019.

Tarrant, P. V., Kenny, F. J. and Harrington, D. 1988. The effect of stocking density during 4-hour transport to slaughter on behaviour, blood constituents and carcass bruising in Friesian steers. *Meat Sci.* 24(3):209–222.

The Congress of the United States. 2018. Transporting Livestock Across America Safely Act. Available at: https://www.congress.gov/bill/115th-congress/senate-bill/2938/text?format=txt. Accessed January 2020.

Tucker, C. B., Mintline, E. M., Banuelos, J., Walker, K. A., Hoar, B., Varga, A., Drake, D. and Weary, D. M. 2014. Pain sensitivity and healing of hot-iron cattle brands. *J. Anim. Sci.* 92(12):5674–5682.

United States Code Annotated. 1994. Title 49. Transportation. Subtitle X. Miscellaneous. Chapter 805. Miscellaneous. § 80502. Transportation of animals (28 hour rule). Available at: https://www.govinfo.gov/content/pkg/USCODE-2011-title49/pdf/USCODE-2011-title49-subtitleX-chap805-sec80502.pdf. Accessed December 2019.

USDA. 2010. Beef 2007–08 Part IV: Reference of Beef Cow-calf Management Practices in the United States, 2007–08. Available at: https://www.aphis.usda.gov/animal_health/nahms/beefcowcalf/downloads/beef0708/Beef0708_dr_PartIV_1.pdf. Accessed December 2019.

USDA. 2018. Dairy 2014, Health and Management Practices on U.S. Dairy Operations, 2014. Report number 3, USDA-APHIS-VS-CEAH-NAMHS, Ft. Collins, CO. Available at: https://www.aphis.usda.gov/animal_health/nahms/dairy/downloads/dairy14/Dairy14_dr_PartIII.pdf. Accessed December 2019.

Woods, J. and Grandin, T. 2008. Fatigue: a major cause of commercial livestock truck accidents. *Vet. Ital.* 44(1):259–262.

Index

www.ingramcontent.com/pod-product-compliance
Ingram Content Group UK Ltd.
Pitfield, Milton Keynes, MK11 3LW, UK
UKHW021826100926
8556IPUK00003B/185

9 781786 764591